ENERGY PROBLEMS AND PROSPECTS

LIST OF MEMBERS ON THE STEERING COMMITTEE

1. Prof. S. Manzoor Alam (Chairman)
2. Prof. M.K. Khera (Convenor)
3. Prof. B.D. Sharma
4. Prof. A. Wahid
5. Dr. H. L. Razdan
6. Sh. G.D. Wadhwa
7. Prof. Majid Husain, and
8. Dr. D.S. Parmar (Secretary).

ENERGY PROBLEMS AND PROSPECTS

Studies on Jammu and Kashmir

Edited by
M.K. KHERA
B.D. SHARMA
MAJID HUSAIN

CONCEPT PUBLISHING COMPANY, NEW DELHI 110059

ISBN 81-7022-035-1

First Published 1991

Published and Printed by
Ashok Kumar Mittal
Concept Publishing Company
A/15-16, Commercial Block, Mohan Garden
NEW DELHI - 110059 (India)

Lasertypeset by
Microtech Advance Printing Systems (P) Ltd.
H-13, Bali Nagar
NEW DELHI - 110015

CONTENTS

SESSION II

CONVENTIONAL SOURCES OF ENERGY IN J & K STATE

SESSION III

PROSPECTS OF NUCLEAR POWER IN J&K STATE

SESSION IV

ALTERNATE SOURCES OF ENERGY IN J & K STATE

SESSION V

ENERGY MANAGEMENT AND CONSERVATION

SESSION VI

OPENING ADDRESSES

CONCLUDING ADDRESSES

FOREWORD

Energy is of vital importance to all of us and has always been so from the dawn of civilisation. As a matter of fact, it is the very basis of existence. This book, the outcome of a seminar, covers the subject of "energy problems and prospects" with particular reference to the State of Jammu and Kashmir. It is hoped that it will help the Government of Jammu and Kashmir in the evolution of a comprehensive energy policy based on its natural resource endowment and geo-climatic situation and consistent with the preservation of its ecological balance.

In the early seventies, when oil prices shot up steeply; sudden realisation dawned on nations most affected that the days of cheap energy were over and they had to identify and develop alternative energy sources much sooner than they expected.

The last one decade has witnessed an explosive growth in energy research both in direct solar energy and in its indirect forms like wind, waves, tide, ocean thermal gradient, biomass etc. Although these are all important in their own rights, those related to our rural setting deserve special consideration.

The energy consumption pattern and adaption of new technologies can deeply affect our social relations, life style and human settlement. Therefore, the need of the hour is to have an imaginative and far-sighted policy of energy production and its management and distribution in order to establish the requisite technological infrastructure consistent with the objectives of our egalitarian society.

Of course, alternative sources of energy are available, but these are either under-developed or elude the norms of commercial exploitability. Solar energy appears to be the most promising energy source for the future as apart from its non-polluting qualities, the amount of energy which is available for conversion is many orders of magnitude greater than all the present world requirements and it is there for all of us to share in the most equitable way possible. The development of safe fusion energy is another future hope of mankind.

The energy situation in the country is far from satisfactory. In the State of Jammu and Kashmir it is near dismal. There is acute scarcity of electricity and the gap between demand and supply is ever-widening. The deficit which was only 64 MWs in the early eighties is likely to go up to the staggering figure of about 500 MWs by the end of the current decade. This acute scarcity of power has led to a massive and reckless exploitation of green forests for firewood with disastrous effects on the ecology and agricultural productivity. Besides, shortage of power, the most important infrastructural input, has caused stagnation of the economy by inhibiting both industrial and agricultural growth.

One important factor responsible for the present situation of acute shortage is the poor management of the supply of power resulting in transmission and pilferage of more than 40 per cent. On the other hand, although the State has no coal, oil or natural gas reserves, its potential for hydro-electric generation of electricity is immense indeed. Already schemes for a generation capacity of 10,000 MWs have been identified and ready to be investigated and taken up for execution if funds for the purpose are available.

The object of this Seminar was to evaluate critically the energy situation in the State, assess its needs and suggest ways and means to minimise and, if possible, eliminate in the foreseeable future the scarcity of power. The seminar is the first of the series of problem-oriented seminars which the University of Kashmir is organizing in collaboration with the State Government and concerned agencies of the Government of India.

I entirely agree with the Editors of this book that the overall objective of the Seminar "Problems of Energy Prospects in the Jammu and Kashmir State" was to bring together on a common platform energy experts from Educational, Research Institutions as also experts from the Government Departments in the Centre and the State as well as policy planners and decision-makers so that a comprehensive policy to tackle the energy problem is evolved.

There was a need for a volume of this kind which describes the present state of knowledge and offers useful information and relevant data relating to various sectors of energy related to this State. I hope this book not only fulfils that need but will also provide fresh impetus for further research, development and policy formulations.

JAGMOHAN
Governor, J & K

PREFACE

Energy is the very basis of our existence and covers the entire gamut of human life. It has played a very important role in the lives of people in any society whether ancient or modern. Per capita consumption of total energy in any state or a country is a measure of its socio-economic development. The State of Jammu and Kashmir has been passing through a period of serious power shortage which has hampered the overall socio-economic progress of its people. There is an urgent need to understand the energy problems in its varied manifestations to safeguard the environment from disastrous ecological disturbance and to evolve an appropriate and viable solutions to the problems of energy. Consequently, the University of Kashmir, in close collaboration with the Atomic Energy Commission and the Government of Jammu and Kashmir, organised a three-day National Seminar on 'Energy Problems and Prospects in the Jammu and Kashmir State' which was held in Srinagar from June 24 to 26, 1985. The seminar covered the following themes in five technical sessions:

I. Energy Requirements of the Jammu and Kashmir State.
II. Conventional Sources of Energy in the Jammu and Kashmir State;
III. Prospects of Nuclear Power in the Jammu and Kashmir State;
IV. Alternate Sources of Energy and their Potential in the Jammu and Kashmir State; and
V. Energy Management and Conservation.

This book brings together various papers presented in the Seminar and comprises its proceedings. The National Seminar turned out to be a unique event as it brought together for the first time on a common platform distinguished energy experts, policy planners, decision-makers, the officers of the Power and Development

Department of the Jammu and Kashmir Government, energy researchers in educational and other institutions and some of those who in one way or the other were concerned with the energy development programmes in the country to tackle the energy problems of the state. The final recommendations of the seminar that emerged after a thorough discussion of the papers and as a result of the interaction amongst participants were read in a special valedictory session with the hope that it would constitute a 'plan of action' for the state government to follow. A summary of the report and the recommendations are given at the end of the book.

Shri Jagmohan while inaugurating the seminar reminded the audience that we hardly realise the extent of damage being caused by the energy crisis. He pointed out that even a 10 per cent power cut results in a loss of Rs. 7000 crores per year in the industrial sector alone and emphasized the need to bring down the transmission losses to overcome the energy crisis. Shri Jagmohan asserted that there could be no greater resource than the inventiveness of the human mind which alone could transform our liabilities into our assets but the innovative spirit was undermined if the social and cultural ethos of a nation generated a sense of indiscipline among its people. He stressed the need for a new cultural revolution not only at the national level but also at the international level. Shri Jagmohan posed a profound question as to who would supply the necessary energy to trigger such a revolution. The Governor also made a couple of suggestions for the seminar to consider and ponder over in its deliberations.

Dr. Raja Ramanna discussed the problems and prospects of power in India and analysed the three options India had chosen in regard to the generation of energy in the country and its relevance to the Jammu and Kashmir State.

The papers in the first technical session give a neat exposition to the energy requirements of the Jammu and Kashmir State. There are six papers in this part which give an insight into the energy scenario of the State at the macro and micro levels.

The first paper by Qazi H.U. Naqash gives a vivid account of earlier history of the subject and highlights the effort the State Government is making in the development of electric power. The paper on the energy requirements for the Jammu & Kashmir State by G.N. Mujoo is a systematic attempt on the problem, emphasizing the sectoral requirement of energy in the State in which sophisticated statistical techniques have been used to ascertain the future projections. Dr. Parmar J. Shankar and Dr. Peer examined the energy requirements of the agriculture sector of the State with a set objective to find out the alternate sources of energy to meet the growing demand. Dr. Nisar Ali analysed the spatial and temporal aspects of energy problems in the State in his paper to explore into the consumption requirements and future projections. The subsequent two papers of this part deal with case studies made by Dr. Husain and Zutshi which pertain to the fuel consumption patterns in the Kandi (hilly) rural areas and the Municipality of Srinagar respectively. Fuelwood as the dominant source of energy has been established empirically by Husain by generating data from the villages of Naranag Catchment; while the fuel consumption pattern in Srinagar Municipality shows wide variations constrained by the socio-economic status of individual families and their income groups.

The second session of the book contains four papers on conventional sources of energy in the Jammu & Kashmir State and Keynote Address by Dr. S. Varadarajan, former Director-General of the Council of Scientific and Industrial Research. Dr. Varadarajan while commending the papers (presented in the seminar) for their quality and thorough analysis of different types of energy that could be developed within the State highlighted a number of interesting and important problems relevant to the development of energy in the State and made thought-provoking suggestions having a bearing on the economy of the State in the future. The studies presented in the papers that follow point out the difficulties that were experienced in implementing various hydel projects especially on a difficult terrain. One of the papers (Y.L. Khular) analyses the hydel potential of the Jammu & Kashmir and makes out a case for comprehensive studies for utilization and exploitation of water resources of the Himalayas for the next three decades. A paper by Mohammad Sultan Bhat and Shah Nawaz Alam highlights the significance of firewood to the State's economy and the need for proper fuel planning.

The third session of the book deals with the prospects of nuclear power in the State. Dr. P.K. Iyanger discusses the fundamentals of the nuclear power production based on fission. Dr. M.R. Srinivasan, former Chairman of the Atomic Energy Commission, presented the scenario of nuclear electricity generation in the world and the Indian Nuclear Power Programme. This section also deals with the safety aspects of nuclear power and its feasibility for the Jammu & Kashmir State. A paper on the Fabrication of Nuclear Components gives an account of the progress achieved in the country. The last paper gives a detailed account of the operational aspects and experience gained in the design, construction, commissioning and operation of nuclear units in India.

The fourth session of the book is devoted to the alternate sources of energy in the Jammu & Kashmir State. Prof. M. Ramakrishna Rao gives an account of solar-thermal energy systems being developed in the country and the contribution these systems could make in providing alternate energy sources. Arun Kumar describes the steps being taken in the Jammu & Kashmir State for harnessing new and renewable sources of energy. Prof. Thapar discusses the relevance of hybrid micro hydel projects for the State. Prof. Bokadia stresses the need to utilise biomass for conversion of energy.

The fifth session of the book deals with the Energy Management and Conservation. Prof. K.L. Chopra in his keynote address stressed the importance of energy conservation and examined the relevance of photovoltaic devices. He also discussed the merits and limitations of NRSC devices in relation to the Jammu & Kashmir State. Unfortunately his keynote address could not be reproduced here. The last paper (Dr. A. Aima) discusses the issues in the energy planning.

The last part of the book gives the proceedings of the inaugural session beginning with a welcome address by Prof. Shah Manzoor Alam, the Vice-Chancellor of the University of Kashmir, followed by the inauguration of the seminar by the Governor of the State, Shri Jagmohan, and a Keynote address by Dr. Raja Ramanna, the Chairman of the Atomic Energy Commission.

The book-one of the first of its kind, should prove useful to the energy experts and energy planners and may provide an additional impetus to the researchers and policy-makers. It is hoped that the book will also enable them to focus their attention on finding suitable solutions to the energy problems of the State. The book should have much wider appeal in the sense that quite a number of problems are common to many other States of the country.

In the end we would express our regrets for any errors and omissions that might have crept in spite of our best efforts in reproducing the talks/comments delivered by some of the experts as these had to be transcribed from the magnetic tapes.

M.K. KHERA
B.D. SHARMA
MAJID HUSAIN

ACKNOWLEDGEMENTS

On behalf of the University of Kashmir and on our behalf we express our grateful thanks to the following agencies for providing the University the financial support for organising the national seminar and bringing out this book: 1. Department of Power & Development, Government of Jammu & Kashmir; 2. University Grants Commission; 3. Ministry of Irrigation & Power, Government of India; 4. Department of Science & Technology, Government of India; 5. Department of Non-conventional Energy Sources, Government of India; and 6. Atomic Energy Commission.

Special thanks are due to the participants who contributed the papers to the seminar and/or actively participated in its deliberations. We are very much indebted to Shri Jagmohan, the Governor of the Jammu & Kashmir State, for providing the inspiration and initiating the first step in organising the seminar. We are also grateful to Prof. Shah Manzoor Alam who was the moving spirit behind the seminar. He was a constant source of strength to us and continually encouraged us in bringing out this book. We acknowledge the support given by the University of Kashmir in providing a subsidy of Rs. 14000 out of the funds received for the seminar to the publishers for bringing out this volume. We acknowledge the untiring efforts of Dr. D.S. Parmar who worked as the Secretary of the Steering Committee and played the major role in helping to organise the seminar.

We highly appreciate the unstinted support given by Dr. H.L. Razdan in the organisation of the seminar. Thanks are also due to Dr. M.A. Peer for the difficult assignments he undertook and the helped he provided in transcribing a number of talks (delivered during the seminar) from the magnetic tapes. We acknowledge the hard work put in by the following Rapporteurs who were kept busy continuously till late hours every evening to complete the report of each technical session; Prof. M.M. Bokadia; Prof. Y. Singh; Prof. Majid Husain; Prof.

Abdul Wahid; Dr. H.L. Razdan; Shri Arun Kumar; Dr. D.S. Parmar; Shri A.M. Mattoo; Shri T.N. Bhat; Dr. M.A. Peer and Dr. Ashok Aima.

Many other persons contributed to the success of the seminar in one way or another and it is not possible to name all. However, we may mention the following: Dr. S.K. Khosa; Shri Mohd. Mustafa; Dr. Nisar Ali; Dr. (Miss) Bimla Labroo; Dr. (Mrs) Atula Khosa; Dr. (Miss) Renu Vakil; Dr. Kuldip Raina; Shri Altaf Ahmed; Shri Anil Ganjoo; Shri Mohammad Hussain; Kumari Indira Kachroo; and Shri Nazir Ahmad Banday.

M.K. KHERA
B.D. SHARMA
MAJID HUSAIN

LIST OF CONTRIBUTORS

Ashok Aima
Department of Commerce-Management Studies, University of Kashmir, Srinagar
Prof. Shah Manzoor Alam
Vice Chancellor, University of Kashmir, Srinagar (1984-87)
Shah Nawaz Alam
Research Fellow, Department of Economics, University of Kashmir, Srinagar
Nisar Ali
Department of Economics, University of Kashmir, Srinagar (Retired)
Mohammad Sultan Bhat
Department of Economics, University of Kashmir, Srinagar
M.M. Bokadia
Professor in Chemistry, University of Kashmir, Srinagar (Retired)
Prof. Maheshwar Dayal
Secretary, Department of Non-conventional Energy Resources, Government of India.
Mondira Dutta
Lecturer, Department of Geography and Regional Development, University of Kashmir, Srinagar
Majid Husain
Professor and Head, Department of Geography and Regional Development, University of Kashmir, Srinagar (presently Deptt. of Geography, Jamia Millia, New Delhi)
P.K. Iyengar
Director, Bhabha Atomic Research Centre, Trombay, Bombay. (presently Chairman, Atomic Energy Commission)

Shri Jagmohan
The Governor, Jammu and Kashmir State (presently member Rajya Sabha)

B.N. Kaul
Chief Engineer, Civil Investigation and Designs, Power Development Department, Jammu and Kashmir

Brigadier Y.L. Khular
Director, North Western Circle, Survey of India, Chandigarh

A. Krishnan
Nuclear Power Board, Bombay

Arun Kumar
IAS, Additional Secretary to Government of Jammu and Kashmir, Power Development Department, Srinagar

G.N. Mujoo
Joint Commissioner (Power), Power Development Department, Jammu and Kashmir, Srinagar

G.V. Nadkarny
Chief Superintendent, Rajasthan Atomic Power Station, PO: Anushakti, Via Kota, Rajasthan

Qazi H.U. Naqash
Development Commissioner, (Power) Power Development Department, Jammu & Kashmir, Srinagar

D.S. Parmar
Department of Physics, University of Kashmir, Srinagar

M.A. Peer
Department of Physics, University of Kashmir, Srinagar

Raja Ramanna
Chairman, Atomic Energy Commission, and Secretary, Department of Atomic Energy (presently Minister of State, Defence, Govt. of India)

M. Ramakrishna Rao
Indian Institute of Science, Bangalore

G.M. Shah
Former Chief Minister, Jammu and Kashmir State

J. Shankar
Director, Centre of Research for Development University of Kashmir, Srinagar (Retired)

M.R. Srinivasan
Former Chairman, Nuclear Power Board, Department of Atomic Energy

O.D. Thapar
Director, Alternate Hydro Energy Centre, University of Roorkee, Roorkee

Prof. S. Varadarajan
The then Director General, Council of Scientific Industrial Research

G.D. Wadhwa

Bhupinder Zutshi
Lecturer, Department of Geography and Regional Development, University of Kashmir, Srinagar.

SESSION I

ENERGY REQUIREMENTS OF JAMMU AND KASHMIR STATE

Chairman: Dr. P.K. Iyengar, Director, Bhabha Atomic Research Centre

ONE

PROBLEMS AND PROSPECTS OF ELECTRIC ENERGY IN THE JAMMU AND KASHMIR STATE

QAZI H.U. NAQASH

Before I make observations on the problems and prospects of electrical energy in the J&K State I would like to give a brief account of the development of power in this State.

It is a well-known fact that Jammu and Kashmir was one of the pioneers in the field of power generation in this country. While the then Mysore State was the first to have a luxury of electric power in the beginning of this century, the Jammu and Kashmir State was not far behind. The old Mohora Power House, which was the only source of power in Kashmir until 1955, was commissioned as early as 1908. Having remained in operation until the year 1961, generating 2.4 MWs by utilising the waters of river Jhelum, diverted at Boniyar into an eight mile-long wooden flume, this station had 4 Nos. 1,000 KVA alternators each coupled to a pelton wheel. While the generation voltage was 2.3 KV stepped up to 30 KV for transmission, its frequency was an odd 25 cycles per second. This power house was initially conceived for operating a drudging plant on River Jhelum at Sopore Town. Besides the City of Srinagar, which obviously was not a large city of those days, the facility of electric supply was later on extended to the then towns of Sopore and Baramulla as well. Although the power plant damaged during the termoil of 1947, was repaired and recommissioned shortly thereafter by the State engineers, its capacity got reduced to less than 2 MWs. Although we were pioneers in this field, it is ironical as well as unfortunate that no significant addition to the power generation was made in this State up to 1955. Thus the total installed capacity of Kashmir until June 1955 was just 2 MWs. Jammu at that time was getting its supply from a small thermal station at Miransaheb and the micro hydel station on Ranbir Canal in Jammu City, besides a few dispersed Diesel Generating sets. Work had, however, been taken up on Sindh Hydro Electric Project of Ganderbal soon after Independence. Conceived for an ultimate capacity of 18 MWs, this project (with initial capacity of 6 MWs only) was commissioned in the year 1955. As this generation was at 50 cycles per second, the Mohora Power with its 25 C/S frequency continued feeding certain areas in isolation.

Soon after the power became available from Sindh Project, the facility was extended to other towns and villages of the Valley. The available capacity however, being limited, the sudden abundance experienced after years of suffering proved very short lived. The demand rose at such a quick rate that in 1959 itself i.e. just four years after commissioning the Ganderbal Project, the Valley again started experiencing shortage during peak load hours. The Government had meanwhile decided to add 9 MWs in Ganderbal and create another 9 MWs capacity in Mohora after scrapping the 25 cycle machines over there which had more than outlived their useful lives. These projects accomplished in 1961 and 1963 respectively raised the total installed capacity of the Valley to 24 MWs. The distribution network having by this time spread extensively and power carried to various areas the additional capacity so created got fully absorbed by 1963 before one could feel its real impact. No doubt the summer would pass comfortably with sufficient power available from the afore-mentioned run-of-the-river stations, yet deficit would be experienced during winter when power was needed more. There was at that time no interconnection between the power systems of Kashmir and Jammu. Jammu Province received about 16 MWs from Punjab, allotted from Bhakhra system, over 66 KV transmission lines. This much power, at that time, was sufficient for loads of Jammu.

With a view to filling the winter deficit, even partially, a 5 MW diesel generating station was set up at Srinagar in 1967 raising the installed capacity to 29 MWs. Work was simultaneously taken up on Upper Sindh Project at Sumbal for a capacity of 22.6 MWs. Constructing a transmission line over Pirpanchal, for connecting the Valley with Jammu and through Jammu with the Northern Grid, was mooted, at this very time. This task, considered impossible earlier, was accomplished by the State engineers in seventies. Thus a Double

Circuit 132 KV Transmission line, crossing Pirpanchal at 10,000 feet ASL came into existence in the year 1975.

The following projects got constructed by then:

	Capacity in MWs.
1. Kalakote Thermal Station	22.5 MW
2. Chenani Power House	25.0 MW
3. U.S. Stage-1	22.5 MW

Work had also been taken in hand on the State's major power station of Lower Jhelum (105 MWs).

Power network in the State was simultaneously spreading gradually all over the State under the R.E. programme launched in 1972. The result is that J&K is perhaps one of the very few States where Rural Electrification has been completed in over 80 per cent villages, 97 per cent in Kashmir and 85 per cent in Jammu region. This extensive spread of distribution network has obviously resulted in an over increasing rise in demand for power.

The total installed capacity of the J&K system by 1979 had risen to 208 MWs. A few micro and mini hydel stations like Poonch (2 x 30 and 1 x 90 KWs), Rajouri (2 x 350 KVA), Baderwah (2 x 350 KVA) and Jammu (2 x 500 KVA) were additionally in isolated operation at that time. Work had also been taken up for reconstructing a 220 KV link between Udhampur and Sarna (Punjab) so that a stable Inter-state connection could be established with the Northern States. This line was commissioned at 132 KV level, initially, in the year 1976.

It can be seen that J&K system with an installed capacity of 208.62 MWs by the year 1979 had grown nearly 100 times since early fifties. More significant is the development of interconnections, not only between Kashmir and Jammu but also between the J&K and the Northern Grid. This arrangement besides stabilising operation of the small power stations on our system provided excellent flexibility in operation. Thus while the State started exporting its surplus energy (through a small quantity) during lean load periods in summer, it could cover its deficits partly, drawing power from the Northern Grid under the arrangement made through good offices of the NREB.

Even though the LJHP had been fully commissioned, the situation became critical again in 1980 when against an estimated energy demand of 870 MUs for 1979-80, the total availability from all sources was only 403 MUs. A loosing rate between the supply and demand continued thereafter as no new generation got created in the State since 1979, except a small share of 10 MWs, corresponding to 65 MUs annual energy, obtained from Bhakra complex. Thus against the estimated demands ranging from 1000 MU to 1600 MUs during the 6th Plan period, the total availability which includes the additional assistance arranged from the Northern Grid varied between 900 and 1300 MU respectively, leaving thus very large gaps especially in winter periods.

Jammu and Kashmir is one of the few States blessed by Nature with an abundant wealth of hydel potential. Preliminary studies have identified hydel power potential of over 13000 MWs. A number of schemes have in this regard been conceived. Out of the major schemes conceived on Chenab, the Salal (345 MWs) at an advanced stage and work has been started also on Dulhasti project, both being executed as Central projects. Under an agreement with Central Government, the Salal Project is to supply a minimum of 80 MWs power corresponding to 35 per cent annual energy i.e. about 800 MUs to the State. Dulhasti power is also to be shared in the region under a National formula giving 20 per cent share to the State. While Salal is scheduled for commissioning in 1986, Dulhasti may come during the next plan. One more major project in the State namely 480 MW Uri is also to be taken up as a Central Project. Having received techno-economic clearance years back, and acknowledged as one of the easiest projects of its size, it is unfortunate that construction has not started so far even though the State Government have since given consent for doing so. It may be revealed that the State Government, gave this consent reluctantly as it looked impossible to fund the project within the State's plan size. It is hoped that the Central Government, with its vast resources, will execute the project more expeditiously.

A 105 MW Hydel Project, the third project utilising Sindh waters has been taken for construction in the State sector. With its power house at Kangan, (District Srinagar), the project will have an installed capacity of 70 MWs in first phase. It is expected to yield 354 MUs annually rising to 535 MUs in the final phase. Its scheduled completion is 1988.

Some additional energy is also expected from Lower Jhelum Project as a result of regulation of water to be made for navigation purposes in the River Jhelum. A Lock under construction down stream of Wular is to release a regulated discharge during winter months which will automatically help the Lower Jhelum Power House to produce more energy during winter.

The State is also to draw its lawful shares from various Central stations in the region. Although for reasons not explained to the full satisfaction of the State, no allocation of share has been made so far from the first phase of Singrauli Super Thermal Power station since gone into production, an assurance has been held out that a share will be allocated instead, from its 2nd phase.

Narora Nuclear project too is to have a power and energy share for the J&K State. Similarly Chemara Project in Himachal is also to yield over 100 MWs to the State according to the National Sharing Formula.

Even though sizable chunks of energy are to become available hopefully in coming years, yet its demand in the State has been rising so fast that it appears wellneigh impossible to bridge the ever widening gap between the demand and the supply. The 12th Load survey by the CEA has shown the demand rising from 385MW/1857MU in 1985-86 to 632MW/3047MU in 1989-90. After accounting for the likely availability from all sources, the deficit left is rising from 169MW/653MU to 368MW/752MU respectively for the years 1985-86 to 1989-90. A gigantic demand of 2328MW/1121 BMU has further been projected up to the year 2000 in the said study.

The question that has been evading an answer is how to cover this ever yawning gap immediately. Perhaps no one has a ready solution. The situation with power availability especially in the Valley has its own peculiarity. While our shortages becomes very acute in winter because generation from our own run-of-the-river-stations fall to nearly 1/3rd their capability, the demand on the other hand increased tremendously. Summer conditions are also not much different any more. Gone are the days when this State used to have surplus energy during summer. That position has changed totally. We are as such in deficit even during summer but the condition in winter, for the reasons given earlier, becomes worse.

Our rivers which are capable of generating thousands of Gegga-Watt hours have characteristic limitations of the run-of-river pattern. As the discharge in snow fed rivers falls far short of requirement in winter, firming up power has to be found for optional utilisation of the waters. A few storage schemes identified for this purpose are no doubt under investigation but these are very long range projects. With a view to fulfil this duty the State had proposed setting up a thermal station of adequate capacity near its rail head at Jammu.

This proposal did not, however, find favour with the Central Government which assured power supply from the Super Thermal Stations instead. But this promise has unfortunately not been fulfilled so far. We have thus been left with the problem without giving us a solution. May I have the views of the speakers here on this subject as well.

For a number of projects having been identified (and a few feasibility reports prepared), the derth of finances mainly comes in the way of our executing some of these. In order to overcome this difficulty partially, the State Government has extended offers to willing industrialists in the country to come forth and implement such projects for their proposed industries within the State. The generating station under this arrangement can even be away from industry, and the power so generated pumped into the State's Grid can be carried and delivered to the unit after recovering certain royalty towards water resource and wheeling charges etc.

Having briefly narrated the history of Power Development in the State and the precarious condition of its vast power network which remains mostly under-fed for want of sufficient generation, I would request the speakers to come forth with practical suggestions for solving this problem in as short a time as possible.

May I also give you briefly insight into the Power Development Activities in Ladakh region. Electrification of the two major towns of Ladakh, Leh and Kargil, was carried out in 1965 by installing small Diesel Generation Stations. The power network was thereafter extended to numerous villages successively under the various R.E programmes. The number of towns and villages presently electrified, mostly from Diesel Generation, in the two districts of Leh and Kargil was 46 and 49 Nos. respectively ending year 1984-85. As the Diesel Generation produces very costly energy especially in distant places of high altitudes and since there are large number of sites for Hydel Generation available in the region, the P.D.D. have prepared a number of Micro and Mini Hydel schemes. Work is also in hand presently on the Stakna Hydel Project of 4 MWs capacity. This project, when commissioned, hopefully in the middle of 1986, will meet the power requirements of Leh town and its surrounding villages to a large extent for a few years to come.

Similarly in Kargil District a mini Hydel Project of nearly 4 MWs is to be constructed on river Suru in the Kargil town itself. Some works on this project have already been taken in hand and it is hoped that a more active construction will begin soon.

Over a dozen micro hydel schemes have additionally been identified and feasibility reports prepared in this District. Diesel generation has also been started in Zanskar Valley for meeting lighting requirements of the population there. This will, however, be replaced with hydel power after the Haftal 1 MW power station gets commissioned. A project report for this scheme has also been prepared.

There is large power potential in the River Indus in Ladakh. Suru, which flows through Kargil, has also sizable power potential. Exploitation of these potentials will however have to wait until power evacuation problem has been overcome. The P.D.D. has under its consideration construction of a pilot 33 KV transmission line between the Valley and Kargil in this regard. When it proves its feasibility, one can think of stepping it up to higher voltages for transmitting larger quantities of power.

Although having fallen short of its consumers expectations, may I, before ending, make a brief mention of the contribution made by the State's Power system in meeting their energy requirements. Our statistics show the energy supplied has risen from 136 MUs in 1965-66 to 1358 MUs i.e. 10 times in 1984-85. On a rough calculation, it has been found that for meeting the cooking requirements alone of the domestic consumers in the urban areas of the State, taking only 50 per cent of such consumers with an average of 1 KW heater for each family, nearly 1800 MUs has gone for this purpose during this period of 20 years. We have thus contributed 40 lakh trees worth energy taking fire-wood content of 30 Cft in each tree, which otherwise had to be felled for meeting this requirement of the people.

Is it not a luxury to use electric power for heating and cooking when it is required more for running industries and for other productive uses? A solution has therefore to be found for meeting their cooking energy requirement of the people from sources other than electricity. This could be one of the quickest solutions to the present problem of shortage of power to the industry and other essential consumers in the State. When I was in China on a short visit in winter 1983, I enquired from the people there whether they used electricity for heating and cooking requirements in their homes. The reply given was that it was a luxury which they certainly could not afford. Their Government, however, provided them sufficient supplies of coal for this purpose. We in this State may have to think on similar lines, and supply coal/kerosene and Gas to the people even if it has to be subsidized to attract them rather than mis-using electricity for this purpose.

TWO

ENERGY REQUIREMENTS FOR JAMMU AND KASHMIR STATE

G. N. MUJOO

INTRODUCTION

Electric power has proved itself to be one of the most powerful instruments of social and economic change that the world has seen. It has many unique properties, e.g. to the household consumer, it represents the most convenient and versatile form of energy providing simultaneously motive power, heat or light. In many industries there is no substitute for electric power. Though its use in the State of Jammu and Kashmir is largely confined to domestic use, yet it is playing an increasingly important role in agricultural and industrial sectors. The rising cost of petroleum products, the growing burden they throw on the country's balance of payments as also their impact on the eco-system of the Valley, necessitate their substitution by power in the transport sector.

Against these advantages, electric power suffers from two serious shortcomings. Firstly, it cannot be stored and must thus be consumed when produced and, secondly, it is the most expensive form of commercial energy in both capital and operating costs. The main problem is how best to go about developing this vitally important but expensive energy source in relation to other sources so that the large socio-economic goals of employment generation, poverty amelioration, provision of minimum needs and attainment of self-reliance within the severe resource constraints may be achieved.

It is clear that electricity by virtue of its being a superior form of energy has emerged now as an essential input, very often in replacement of other forms of energy. In absolute terms, however, the level of electrical energy consumption in Jammu and Kashmir State is much below the national average and far behind that of the developed countries. The per capita electricity consumption of Jammu and Kashmir in 1960-61, 1977-78 and 1980-81 was 14 units, 65 units and 93 units respectively against the corresponding national figures of 38, 121 and 133 units. This has no comparison to 8,487 units (1976) in the USA and 4,166 units (1976) in the U.K.

The growth in demand of power and inadequate capacity addition has resulted in power shortages. These shortages which were sporadic in the early seventies have become serious and sustained now. Power shortages have affected all sectors of economy. In order to understand the reason for the continuing imbalance between demand and supply, several issues have to be examined. Planning of capacity addition and whether the planned addition is realistic to meet the demand is one among them. Forecast of demand for power, both in terms of peak demand and energy, is the first step in the power planning process. With projects having longer gestation periods of 10 years and more the need for long-term demand forecasting covering periods of 15-20 years needs justification. Such forecasts ought to be prepared and updated periodically to form a framework for formulating plans. A time frame of this order will permit long gestation hydel and nuclear projects to be planned along with shorter gestation thermal projects so that the generation mix is optimized and resources mobilized.

Demand forecasts for power are made by different bodies, the Annual Power Survey Committees, the Planning Commission and the individual States. All use different methodology and hence arrive at different figures of demand.

Annual Power Surveys

The Department of Power, Government of India, organizes annual power surveys at regular intervals. These were started in the early sixties on the pattern of semi-annual power surveys of the Edison Electric Institute of the USA. Intended to be carried out annually they were to project the demands for power over a five-year period in order to provide the data for determining measures required to be taken to ensure adequate power availability in a given time frame.

The surveys are conducted under the aegis of a Power Survey

Committee. The Power Survey Directorate of the C.E.A. acts as the Secretariat for the committee. Since 1962 twelve surveys have been carried out and the twelfth Survey has recently been published (March 1985). The results of these surveys are used as one of the inputs into capacity planning and investment in the power sector by the Government.

The methodology adopted for forecasting the medium-term demand in the annual power surveys has been modified from time to time. Various techniques like time series analysis, end-use method, econometric modelling have been tried and each time the Committee presented the forecast, it highlighted the shortcomings in previous approaches also. A modified technique called partial econometric model has been used in the 12th Annual Survey. It is a combination of all past trends, economic growth and end-use methods.

Planning Commission Projections

The Planning Commission works out independently the electricity requirement for the five-year plan period using an input-output model, detailed material balance studies and the capital output ratios. The resource constraints thus get built into its forecast. Because they work to physical targets based on expected rates of growth of the economy which is lower than what is implied in the demand surveys, the demand forecasts of the Planning Commission tend to be lower than those projected by the Annual Power Surveys (APS).

The working group of energy policy, set up by the Planning Commission, in their report published in 1979, has suggested "Scenario" approach in perspective planning with the realization that there will be severe constraints on the availability of oil and coal in future and alternative energy sources have to be developed along with reduction in the use of imported oil. The committee has examined the energy consumption trends in each sector and projected the future trend for two Scenarios–one called the Reference Level Forecast (RLF) and the other the Optimal Level Forecast (OLF). RLF represents the likely level of demand for energy, if the present trends of energy consumption continue and no deliberate measures are initiated to manage the energy supply and demand. The OLF Scenario takes into account the various constraints and assumes certain deliberate policy changes to enable some rate of economic growth with a substantially lower level energy input.

The forecast with regard to the rate of growth in consumption of electricity is shown in the following table. The energy GDP elasticities implied in the forecast are also shown. The rates are average annual compounded percentages. But setting up of a less energy-intensive economy requires major policy changes at the Centre as well as State levels.

	RLF	OLF	Energy/GDP (RLF)	Energy/GDP (OLF)
1975-82	10	10	1.34	1.34
1982-87	8.3	6.2	1.22	0.93
1987-92	8.1	6.8	1.18	0.90
1992-2000	6.6	6.4	1.08	0.95

Various Methodologies

Time Series Analysis

The evolution of a mathematical model to fit in the changes in energy consumption in each sector of activity is time series analysis. It does not, nor has it any provision to, accommodate any special efforts or measure to boost or check the consumption at any point of time.

End Use Technique

Equipped with the historical development or the past trend of energy consumption in any sector of activity and the targets for future development, the course of consumption is evaluated by what is known as end use techniques.

Econometric modelling is the forecast technique wherein various economic parameters like GDP, productivity and their growth rates, etc. form dimensions of the energy consumption.

These techniques have been used till now by various institutions in doing the forecast exercise. The Planning Commission forecasts are investment originated and are necessarily on the lower side because of resource constraints. The C.E.A. forecasts too are not realistic for the State of Jammu and Kashmir where we have to cater to the bare minimum needs of people first and then industrial and other productive sectors. As can be imagined, the suppressed growth and stipulated energy GDP elasticities may be not strictly applicable to the non-industrial sector in Jammu and Kashmir. While the specific consumption increases with the passage of time, increase due to the substitution of other kinds of energy for the preferred electricity has found no place in these forecasts. The State's forecasts in general and of Jammu and Kashmir in particular are not merely aimed at raising more of investment but also at improving the real picture of demand and supply in the near future.

The C.E.A.'s forecasts for the Seventh Plan will not differ much from those of the State's due to the following reasons:

(i) Longer gestation period of the hydel projects.
(ii) Suppressed demand because of the serious and sustained power shortages anticipated.

As such the short-term forecasts of annual power surveys up to 1989-90 are reproduced below to form the States' forecasts as well:

Forecasts – 1985-86 to 1989-90

While preparing the draft for the Seventh Plan "Demand for Power", the planning wing of the C.E.A. adopted a methodology which is a combination of economic and time series modelling. The C.E.A. had suggested a tapered growth rate starting from 10.5 per cent and going down to about 9.1 per cent by the nineties. The projections made in the draft for the Seventh Plan were found to be much lower than those projected in the 11th Annual Power Survey of C.E.A. (itself), possible because of the methodology adopted. The department took up the issue with the C.E.A. as a result of which the projections for the Seventh Plan were revised and are reproduced below:

	Energy requirement MWs	Peak Demand MUs
1985-86	1857	385
1986-87	2104	437
1987-88	2390	496
1988-89	2697	560
1989-90	3047	632

These projections have been made after assessing the growth in the demand under the following categories:

1. Domestic
2. Commercial and Miscellaneous
3. Public Lighting
4. Public Water Works
5. Irrigation
6. Industries
7. Bulk supply to the non-industrial consumers.

In this paper an attempt has been made to forecast the figures beyond 1989-90 up to the turn of the century. The demands/requirements have been assessed sector-wise. The sectoral demand development has been evaluated with due care to the possible substitution of various forms of energy. The methodology adopted is described below:

Domestic

The consumption/requirement in this sector has been estimated on the basis of the following :

(a) the number of consumers
(b) their specific consumption
(c) the substitution of other kinds of energy by electricity.

A growth rate of 12 per cent for the period 1984-85 to 1989-90 has been assumed which takes care of population growth, rural electrification and family size, etc. For the period beyond 1989-90 this growth rate will be arrested for the following reasons:

(i) Rural electrification will be almost complete by then,
(ii) Population growth shall be contained by family planning,
(iii) The rate of break-up of larger families into smaller families will no longer be more than the population growth rate by the end of century, and
(iv) The socio-economic change will be steadier.

A tapered growth rate has been assumed for the period coming down from 12 per cent in 1989-90 to 2.6 per cent in the year 2000-1. The transition also envisages reduction in the number of souls per connection to a reasonable number by the end of the century.

It is most likely that the energy usage will, among other things, follow the GDP. The growth rate of GDP anticipated for the whole of India is given below:

	Scenario-I	Scenario-II
1980-90	6.5%	7.5%
1990-2000	6.00%	8.0%

The growth of specific consumption in the domestic sector as assumed in the 12th APS is nearly the same as of the GDP. It seems that no other factor like Energy GDP elasticities, substitution of other kinds of energy have been considered. Tentative estimates indicate that of the total requirements an urban household uses about 80 per cent of the energy other than electricity and the rural consumers as much as about 95 per cent. This means that out of the total average energy requirements only 8 per cent is met by electricity. The rest is met from wood and its derivatives, a prime cause for deforestation in the State, costlier commodities like kerosene oil, diesel, gas and other kinds used less efficiently. For obvious reasons, therefore, increased dependence on wood and its derivatives and use of petroleum products should be discouraged and a gradual substitution by electricity made. This will be possible only when curtailment-free and reliable power supply is ensured. Thus beyond 1989-90 substitution by 0.5 per cent every year shall be made and the specific energy requirement raised accordingly. It is possible any time to substitute the use of other kinds of energy in the household sector subject to the availability of electricity. Requirement so arrived at is given in Annexure-I.

Industrial Sector

Jammu and Kashmir has a very small industrial set-up at present. It is one of the least industrially developed States of the country. We have a great scope for large, medium and small-scale industries including handicrafts and handlooms.

The reasons for its being industrially underdeveloped are:

(i) Lack of reliable transport,
(ii) Absence of reliable power supplies,
(iii) Lack of marketing facilities, and
(iv) Other special and economic factors.

The State Government, therefore, is extending every possible help to the prospective entrepreneurs for setting up industries in the State. And the response is encouraging. There has been an increase of 20 per cent in the energy sales to this sector in 1982-83 to 1983-84.

In the 12th APS a growth of 3.5 per cent to 12 per cent for the energy requirements for this sector has been adopted with the realization that the State has a large scope for industrial development and that the industries only can arrest the growing unemployment, the growth rate shall get intensified with 12 per cent in 1989-90 to 20 per cent in 2000-1. Thus the requirements would be :

Year	Growth Rate	Energy Requirement
1990-91	13%	502.8-MV
1991-92	13.7	571.7
1992-93	14.4	654.0
1993-94	15.1	752.0
1994-95	15.8	871.8
1995-96	16.5	1015.6
1996-97	17.2	1190.3
1997-98	18.0	1404.5
1998-99	18.8	1668.6
1999-2000	19.5	1994.0
2000-2001	20.0	2392.8

Commercial and Miscellaneous

This sector includes supplies to markets, cinemas, hotels, offices, etc. This will grow at the rate of growth of national economy and progress towards civic welfare. The increase in the tourist trade will also have a direct bearing on the requirement on this front.

In spite of drastic power cuts and shortages the department registered an increase of 15.73 per cent in the energy sales to this sector in 1983-84 as compared to 1982-83. The anticipated requirement would grow at the rate of population growth and GDP growth. The growth rate of 10 per cent for the number of consumers in this sector has been assumed up to 1989-90. Beyond this year the growth rate has been assumed up as 10.8 per cent on the basis of population and economic growth. The specific consumption has been increased by 50 units every year up to 1989-90 and beyond this year is expected to increase at least at the rate of the increase in the purchasing power of general public and the marketability. The requirement is evaluated as under:

Year	No. of Consumers	Specific Consumption (Kwhr)	Energy Requirement (MUs)
1989-90	88578	1700	150.6
1990-91	97436	1800	175.3
1991-92	107179	1910	204.7
1992-93	117897	2024	238.6
1993-94	129687	2146	278.3
1994-95	142655	2275	324.5
1995-96	156921	2411	378.3
1996-97	172613	2556	441.2
1997-98	189874	2710	514.5
1998-99	205064	2872	588.9
1999-2000	221470	3044	674.1
2000-2001	239187	3220	770.2
	GR - 10.8%	GR - 6%	

Irrigation and Agriculture

By the end of the Seventh Plan the agricultural demand by way of various lift irrigation schemes is estimated around 50 MWs which is expected to double itself by 2000-2001 when the demand incident on the system will be around 100 MWs. Most of the lift irrigation schemes have to operate in Jammu and Kashmir where these have to operate in two seasons every year and hence the energy requirement would be more.

The number of pump-sets has been increased by an annual step up of 200 which growth shall be maintained even beyond 1989-90. The requirement evaluation is as under:

Public Health and Drinking Water

The energy demands on this front changes because of the following reasons:

(i) Population growth,
(ii) Sophisticated life, and
(iii) Expansion in drinking water distribution network.

	No. of Pump Sets	Energy Requirements		Total Requirements (MUs)
		Pump Sets (MUs)	LI Scheme	
1989-90	2350	61.2	111.3	172.5
1990-91	2550	66.4	122.4	188.8
1991-92	2750	71.6	134.6	205.2
1992-93	2950	76.8	148.1	224.9
1993-94	3150	82.0	163.0	245.0
1994-95	3350	87.2	179.2	266.4
1995-96	3550	92.4	197.2	289.6
1996-97	3750	97.6	216.9	314.5
1997-98	3950	102.8	238.6	341.4
1998-99	4150	108.0	262.4	370.4
1999-2000	4350	113.3	288.7	402.0
2000-2001	4550	118.5	317.5	436.0
			GR – 10%	

Public Lighting

The actual sales to this sector has been 11.00 MUs in 1983-84 and is expected to be of the order of 25 MUs in 1989-90 as envisaged in the 12th APS. A growth rate of 12 per cent for connected load has been assumed since 1983-84 to cover the various dimensions of growth, viz., (a) Population increase; (b) Better civic facilities envisaged in future plans, etc.

A growth rate of 12 per cent in the connected load of these installations has been taken in the 12th APS. The same growth rate has been maintained and the energy required evaluated as under:

Year	Connected Load (Kws)	Kwhrs/Kw	Energy Consumption (Mw)
1989-90	30566	4500	137.5
1990-91	34234	"	154.0
1991-92	38342	"	172.5
1992-93	42943	"	193.2
1993-94	48096	"	216.3
1994-95	53868	"	242.3
1995-96	60332	"	271.4
1996-97	67572	"	304.0
1997-98	75680	"	340.4
1998-99	84762	"	381.3
1999-2000	94933	"	427.0
2000-2001	106325	"	478.3

The energy consumption for this activity is being evaluated as under:

Year	Connected Load (KWs)	Units/Kw of Connected Load	Energy Consumption (MUs)
1989-90	6711	3500	23.49
1990-91	7516	3550	26.68
1991-92	8418	"	29.88
1992-93	9428	"	33.47
1993-94	10560	"	37.48
1994-95	11827	"	41.98
1995-96	14246	3600	47.68
1996-97	14836	"	53.40
1997-98	16616	"	59.81
1998-99	18610	"	67.00
1999-2000	20843	"	75.03
2000-2001	23344	"	84.04

Supplies to Bulk Consumers

Expansion of defence network in the border State of Jammu and Kashmir, execution of bigger projects and other construction works reflects growing power demand on the system. Although the growth of energy sales in this sector of activity has not been uniform possibly because of large-scale curtailment, etc., growth rate of 10 per cent has been assumed in the 12th APS and the same growth rate is expected up to 2000-2001. The energy requirement for the sector is evaluated as under:

Year	Energy Requirement (MUs)
1989-90	262.0
1990-91	288.2
1991-92	317.0
1992-93	348.7
1993-94	383.6
1994-95	421.9
1995-96	464.1
1996-97	510.5
1997-98	561.6
1998-99	617.8
1999-2000	679.6
2000-2001	747.5

Transport Services

The energy balance statement of the Union of India reveals that energy equivalent to about 1,000 units is used per capita per annum in the transport sector. This consumption has been maintained right from 1971. With this specific consumption, the energy consumed in Jammu and Kashmir in the transport sector might have been 4.80 Twhrs in 1971, 5.496 Twhrs in 1976 and 6.192 Twhrs in 1981. In Jammu and Kashmir electricity has found no place in the transport sector nor has it been included in the Seventh Plan for power demand. The Commission in the Seventh Plan is justified by the long gestation period of substitution. Most of the energy is being supplied by fuels like diesel and petrol, coal, etc. The international crises for this kind of energy, its spiraling price, etc., call for speedier substitution of these non-renewable and costlier sources by electricity. We have a very big scope for expansion of the transport sector which implies that the per capita consumption is also expected to grow. Thus apart from population growth, the energy requirement may grow at the rate of 2 per cent per annum in this sector.

The meagre substitution of half per cent per annum by electricity will probably include proposed services like tram-ways for cities of Srinagar and Jammu and other goods transport service. The requirement incident on the system has been evaluated as under:

Year	Population (millions)	Total Energy for Transport Sector (Twhrs)	Percentage Use of Electricity	Electrical Energy Requirements (MUs)
1990-91	6.103	6.103	1.0	61.03
1991-92	6.255	6.380	1.5	95.07
1992-93	6.412	6.671	2.0	133.42
1993-94	6.572	6.974	2.5	174.35
1994-95	6.736	6.291	3.0	218.73
1995-96	6.905	6.623	3.5	235.03
1996-97	7.077	7.970	4.0	318.08
1997-98	7.254	8.332	4.5	375.00
1998-99	7.435	8.711	5.0	435.55
1999-2000	7.621	9.107	5.51	500.09
2000-2001	7.812	9.522	6.0	571.32

System Losses

The State of Jammu and Kashmir is among the few States having maximum energy losses. It consists of both transmission and distribution (T&D) and commercial losses. The loss figures for the past few years are as under:

Year	1980-81	1981-82	1982-83	1983-84
Loss %	51.80	46.65	44.83	43

Factors which undo the improvement effort in containing loss figures are:

- Expansion in the transmission and distribution system.
- Lack of adequate resources to improve and remodel T&D system.
- Lack of stringent tariff.

— Unplanned urban as well as rural development, thereby resulting in low load densities, and
— other socio-economic considerations.

The C.E.A. has anticipated an annual loss curtailment of 40 per cent in 1989-90. Assuming an improval loss containment rate, the loss figures beyond 1989-90 up to the end of the century may follow the following trend.

Year	1990-91	1991-92	1992-93	1993-94	1994-95	1995-96
Loss %	37	34	31	28	25	23
contd.	1996-97	1997-98	1998-99	1999-2000	2000-01	
	21	19	17	16	15	

Load Factor

The load factor depends upon the mix of various categories of activity and as such may vary from time to time. In the case of Jammu and Kashmir, it may not change so much because of the larger percentage of services are in sectors like household, commercial and public lighting. The load factor of 55 per cent is retained.

Peak Demand

Based on the above the forecast of energy and peak demand up to 2001 is as under:

Present Status

Having forecast the demand of power and energy it is pertinent to take stock of the existing position of power availability in the State and efforts being made to reduce the shortfall in the power supply within the available resources.

The availability of power in the State of Jammu and Kashmir is from the following generating stations:

Energy/Peak Demand Forecast (1989-90 up to 2000-01)

Years	Energy Require-ments at the Consumers and MW	% Loss	Energy at the Gene-rator (MUs)	Load factor %	Peak Demand (MWs)
1989-90	1828.00	40	2047.0	65	632
1990-91	2409.61	37	3824.77	"	794
1991-92	2753.08	34	4171.3	"	866
1992-93	3138.79	31	4548.9	"	944
1993-94	3585.03	28	4979.2	"	1034
1994-95	4101.71	25	5468.9	"	1135
1995-96	4638.08	23	6024.7	"	1250
1996-97	4110.34	21	6721.9	"	1395
1997-98	6039.21	19	7455.8	"	1548
1998-99	7149.69	17	8614.1	"	1788
1999-2000	8176.33	16	9733.7	"	2020
2000-2001	9341.64	15	10990.1	"	2281

Note: The sectorwise break-up is given in Annexure II.

	Installed Capacity MW	Effective Capacity MW	
(a) Kashmir		Summer	Winter
1. L.J.H.P.	105.0	105	65
2. U.S.H.P. Stage-I	22.6	22	17
3. Ganderbal	15.0	12	10.5
4. Mohra	9.0	7	7
5. Bemina & other Diseols	6.0	—	4.5
Total	157.6	146.0	104.0
(b) Jammu			
1. Chenani HEP	25.0	20	12
2. Kalakote Thermal	21.4	—	4
3. Other small stations	3.0	2	1
	49.4	22	17
Grand Total	208		
(c) Imports			
State share from BBMB	—	27	27
Total	208		

The energy available from the above source is 956 MUs per year.

As a result of the completion of ongoing schemes of the Sixth Plan it is expected that additional generation will be available from the projects as given below:

1985-86	956 MUs	No addition expected
1986-87	1233 MUs	(a) Addition of 20 MW from Ningli Navigation lock (b) 35 MW from Salal HEP (c) 8 MW from renovation of Ganderbal and Mohra (40+216+21MUs)
1987-88	1679 MUs	Addition of (a) 70 MWs of 2nd and 3rd units of Salal HEP (432 MUs) (b) 2 MWs Karnah HEP (14 MUs)
1988-89	1875 MUs	(a) Addition of Ist Unit (35 MW) USHP Stage II (177 MUs) (b) Addition of 3 MW of Pahalgam (19 MUs). 2nd Unit of USHP-II (35 MW)
1990-91	2052 MUs	No addition envisaged unless adequate funds are provided to complete the new schemes proposed to be taken up, viz. Sewa III, etc.

In addition to the above the following projects are also likely to be started in the Seventh and Eighth Plans which, when completed, will add to the generation capability of the State.

Medium Hydel Projects	Installed Capacity proposed MWs
1. Lower Kalnai	60
2. Nunwan Batkote	22.6
3. Shitkari Kulan	84
4. L.J.H.P. 4th Unit	35
5. Sewa HEP Stage-III	90
6. Chenani Stage II & III	6
7. Pahalgam	3
8. Athwatoo	7.3
9. Gurez (Asthan Nallah)	1
10. Remodelling Rajouri Power House	3
11. Hirapur	6
12. Boniyar	4
13. Brenwar	2
14. Laripur	2
15. Wangat	2
16. Tangmarg	2

The Government is also contemplating installing diesel generating stations of a capacity of 40 MWs.

All these efforts are, however, insufficient to meet the demand and contain the shortages. It would have been worth while to take up the Uri Project with an installed capacity of 480 MWs in the State sector and achieve its completion earliest possible to mitigate the shortfall considerably. But because of the financial constraints it has not been considered possible to take up this project in the State sector. As such the Government has already taken action to request the Central Government to take up this project immediately in the Central sector. Apart from Salal HEP which is nearing completion one more hydro-electric project, viz. Dul Husti with an installed capacity of 390 MWs has already been taken up by N.H.P.C.

The Department is also trying to develop the transmission network of the State to cater to the load growth. Construction of new lines, extension and improvement over the existing ones are all aimed at ensuring better services. The State of Jammu and Kashmir is connected to the Northern Grid by two 220 KV S/C lines, one between Sarna and Udhampur and the other between Sarna and Gladni to be extended to Salal H.E.P. These two lines have a carrying capacity of 243 MWs only. Further the regions of Jammu and Kashmir are connected presently by a D/C 132 KV line between Udhampur and Bemina (tapped at various intermediate points) which has a carrying capacity of 64 MWs only. N.H.P.C. is also constructing one more D/C 400 KV line between Udhampur and Srinagar which is expected to be completed by the end of the Seventh Plan. Studies have been made to improve the power transmission capacity of the 132 KV D/C line. Shunt compensation at various load centres forms part of this effort. This is necessitated because Kashmir will not be able to immediately avail itself of the benefits of the Salal HEP unless we have an adequate transmission facility from across the hills to the Valley.

The improvement in the transmission network will also be accompanied by an increase in the transformation capacities of various grid stations. As against 403 KVA capacity in 1984-85 (Jammu 174 MVA, Kashmir 229 MVA) the capacity will be increased to 951 MVA (Jammu 351 MVA, Kashmir 600 MVA). Similarly the sub-transmission capacity is proposed to be increased from 780 MVA to 1140 MVA by the end of the Seventh Plan.

Acknowledgement

In preparing this note Shri P.N. Tickoo, S.E. Planning, Designs, Srinagar, and Shri M.A. Salroo, Asstt. Engineer, have contributed considerably.

ANNEXURE I

Household Sector Electricity Requirement

Year	Population (millions) Growth Rate 2.6%	No. of Connections (growth rate w.r. to previous year)		Approx. No. of Souls Per Connection	Total Energy Requirements in Each H/H	Use of Electricity in Each H/H Unit	Average % of Electricity Use	Total Requirements in H/Sector
1989-90	5.954	12 %	707402	8(8.4)	10500 Kwhr	900	8.57	636.7 MUs
1990-91	6.103	11	785216	8(7.77)	11130 "	1000	9.0	1012.8 "
1991-92	6.255	10	863737	7(7.24)	11800 "	1120	9.5	1156.4 "
1992-93	6.412	9	941474	7(6.81)	12500 "	1250	10.0	1312.5 "
1993-94	6.572	8	1016792	6(6.46)	13250 "	1464	11.0	1497.2 "
1994-95	6.736	7	1087967	6(6.19)	14050 "	1560	11.1	1937.0 "
1995-96	6.905	6	1158245	6(5.98)	14900 "	1713	11.5	1937.0 "
1996-97	7.077	51	1212061	6(5.48)	15780 "	1893	12.00	2177.64 "
1997-98	7.254	4.3	1264179	6(5.738)	16730 "	2091	12.5	2442.6 "
1998-99	7.435	3.6	1309690	6(5.677)	17740 "	2306	13.0	3020.14 "
1999-2000	7.621	3.0	1348980	6(5.649)	18800 "	2538	18.5	3423.7 "
2000-2001	7.812	2.6	1384054	6(5.644)	19932 "	2790	14.0	3861.5 "

ANNEXURE II

Sectorwise Energy Requirement at Consumers End (MV)

Year	Domestic MUs	Industrial	Commer. & Misc.	Irrigation & Agricul-tural	Public Health & Water Works	Public Light-ing	Bulk	Transport	Total
1989-90	637.0	445.0	151.0	172.0	138.0	23.00	262.0	-	1828.00
1990-91	1012.8	502.6	175.3	188.8	154.0	26.68	288.2	61.03	2409.61
1991-92	1156.4	571.7	204.7	205.2	172.5	29.88	317.0	95.07	2753.08
1992-93	1312.5	654.0	238.6	424.9	192.2	33.47	348.7	133.42	3138.79
1993-94	1497.2	752.8	278.30	245.0	216.3	37.48	383.6	174.25	3585.03
1994-95	1714.1	871.8	234.5	266.4	242.3	41.98	421.9	218.73	4101.71
1995-96	1937.0	1015.6	378.3	289.6	271.5	47.68	464.1	235.03	4638.98
1996-97	2177.64	1190.3	441.2	314.5	304.0	53.40	510.5	318.08	5310.34
1997-98	2442.0	1404.5	574.5	341.4	340.4	59.81	561.6	374.00	6039.21
1998-99	3020.14	1668.6	588.9	370.4	381.3	67.00	617.8	435.55	7149.69
1999-2000	3423.7	1994.0	674.1	402.0	427.0	75.03	679.6	500.09	8176.33
2000-2001	3861.5	2392.8	770.2	426.0	478.3	84.04	747.05	571.03	9341.64

REFERENCES

1. Rajahyaksha, V.G., *Report of the Committee on Power, Ministry of Energy*, 1980, New Delhi.
2. Reddy, K.C., "Load Forecasting and Problems of Demand Management", *I.E. (I) Bulletin*, Vol. 32, Nov. 1982, p. 15.
3. Ramachandra, S.G., "The Nation's Development on Perspective" *I.E. (I) Bulletin*, Vol. 29, p.19, April 1980.
4. India—The Energy Scheme (The presentation by the Indian National Committee at 12th WEC, New Delhi, 1983), *I.E. (I) Bulletin*, Vol. 33, p.1, Oct. 1983.
5. *Eleventh Annual Electric Power Survey of India*, Sept. 1981.
6. *Forecast of Power Demands up to 1990-91, APSC, CEA*, Nov. 1983.
7. *Twelfth Electric Power Survey of India*, March, 1985.
8. *Census of India* 1981 (Part I of 1981 Supplement).
9. *Introduction to Operational Research*, McGraw Hill Publications, New York.

THREE

RENEWABLE ENERGY RESOURCES AND REQUIREMENTS IN J & K STATE

D.S. PARMAR, J. SHANKAR & M.A. PEER

On the basis of preliminary energy studies conducted in Ladakh in 1980, it was concluded that the whole of J & K State is one of the richest in renewable energies and there exists a tremendous potential for energy research both because renewable energy sources are available in plenty, and also because of the demand of the local population for alternative sources. It was therefore, decided to fully understand and explore the locally available energy sources in the State in order that the local population may make the best use of it and thus reshape their economies and energy consumption practices. A project on "Survey of Renewable Energy Resources and Requirements in J & K State" was sanctioned by the Commission on Alternate Sources of Energy, Department of Science and Technology, Government of India, to Professor Jagdish Shankar as Chief Coordinator and Dr. D.S. Parmar as Coordinator. The aim of the project is to study and assess the actual state of renewable energy resources and energy requirements in J &K State with the view that whatever renewable energy technology is adopted, it is adequate, technically and economically sound, culturally acceptable and responds to a real need within the community. In order that the renewable energy technologies become successful as an element in the over all development process of J&K State, the project envisages:

(i) an assessment of the present and future energy demands,
(ii) an assessment of the energy consumption patterns of the

population,

(iii) an assessment of the renewable energy potential in terms of its suitability for different applications,

(iv) knowing the energy demand and its availability in a particular area, to make a comprehensive classification of the demand according to the type of energy required,

(v) to establish relationship between energy demand and renewable energy availability. It can then be determined which energy technologies not only have the greatest potential to succeed but which can also satisfy the greatest demand,

(vi) an assessment of where and how the renewable energies are currently being utilized in the State,

(vii) an assessment of how renewable energy applications can improve the economy of the population, and

(viii) an assessment of how the current energy utilization habits can be improved for better efficiency and what utilization habits should been encouraged for the future.

All districts and tehsils have been covered under the survey. Seven per cent of the villages in each tehsil and 7 per cent of the households in these sample villages have been surveyed. According to these calculations, around 500 villages and about 3,500 households have been surveyed.

The survey team of 10 investigators was accordingly divided into two groups. One of the groups was assigned the task of collecting data and information from the relevant Central Government, State Government, private departments regarding energy availability and its consumption in the State. The other group was assigned the field work to complete the targeted survey schedules.

The survey data has been analysed for assessing the energy consumption patterns of the population. This report gives the preliminary results of the survey.

ENERGY CONSUMPTION PATTERNS

The main emphasis for determination of energy consumption patterns has been on the following operations:

(i) Energy requirements for household operations,

(ii) Energy requirements for agriculture, horticulture and

sericulture,

(iii) Energy requirements for transport, and

(iv) Energy requirements for industry.

The energy consumption patterns for different household operations have been determined for the State population in respect of the following operations:

Preparation of meals, Meals carryied to farm, Fetching of water, Rice cleaning, Making flour, Milk processing, Washing clothes, Vegetable peeling, Ironing, Mud washing, Drying of grains, Others (miscellaneous)

The average energy required in man-hrs/month/person in the State has been given in Table 3.1, where the coal equivalent of this energy for each operation is also given. In this conversion 1 hr of human energy = 3.5×10^{-3} kg of coal calculated on the basis of 3.9 M Col/day dietary human energy = 8 hrs of work and 1 kg coal = 32.6 MJ. The total energy required per month/person for household operations is about .15 kg of coal equivalent or 1.44 KW of electricity. Although the objective of these presentations is not to give the economic analysis of the energy consumption practices the correlations of the energy consumption patterns with education, economy and occupation of a person are clearly visible. The average energy/person/month for rural and urban sectors of the population has similarly been determined, as given in Tables 3.5 & 3.6.

The energy requirements for household operations for different districts on the basis of their population and the actual requirements are given below (on the basis of 1982 consumption levels). The electricity required for household operations is about 9.56×10^{7} KW/year. A rough estimate for 30 per cent efficiency of an electric machine to replace these mechanical operations is 37.55×16^{7}.

If the consumer growth at the uniform rate of 3 per cent for the period 1984-1990 is assumed taking into account all measures for growth control including the population, the following year-wise energy consumption for household operation will be shown in Table 3.2.

Therefore by the turn of this century, the energy consumption just for mechanized household operations is expected to be 32.3 MWh^{-1}.

TABLE 3.1: **District-wise Energy Consumption for Household Operations**

District	Average Energy Man-hrs/ month x 10^{-6}	Average Energy/month Eqv. Coal (1kg) x 10^{-3}	Average Energy/month in KWh of Electricity (KWhr x 10^{-3}
Anantnag	27.0	96.12	865.08
Pulwama	16.7	59.45	535.05
Srinagar	29.8	106.08	954.72
Badgam	14.5	51.62	464.58
Baramulla	27.6	98.26	884.34
Kupwara	13.5	48.06	432.54
Kargil	2.6	9.25	83.25
Leh	2.8	9.97	89.73
Doda	17.5	62.30	560.70
Udhampur	18.7	66.57	599.04
Kathua	15.1	53.76	483.84
Jammu	38.5	137.06	1233.54
Rajouri	12.4	44.14	397.26
Poonch	9.3	33.10	297.90
Total J & K	248.4	885.73	7971.51

TABLE 3.2: **Estimated Year-wise Energy Consumption for Household Op:rations for the State**

Year	Average energy Eq. Coal/ per year (kg x 10^{+6})
1986-87	11.26
1987-88	11.58
1988-89	11.90
1989-90	12.22
1990-91	12.54
1991-92	12.86
1992-93	13.17
1993-94	13.50
1994-95	13.82
1995-96	14.13
1996-97	14.45
1997-98	14.77
1998-99	15.08
1999-2000	15.40
2000-2001	15.72

Fuel Consumption for Household Operation

The fuel consumption patterns for the household operations like cooking, heating water, ironing, lighting and heating of the house have been determined. The various types of fuels in use have been divided into two categories: (a) Commercial, (b) Non-commercial.

Commercial Fuels

The commercial fuels have been considered as those which the consumer buys from the market. These fuels include: firewood, husk, dung, kerosene oil, diesel, electricity, cooking gas, hard coke, soft coke, briquettes, match boxes, petrol and others.

The average consumption of commercial fuels in different forms per person per month are given in Table 3.7. The commercial fuel consumption patterns for rural and urban populations have similarly been given in Tables 3.8 and 3.9. It may be noted from Table 3.7 that main commercial fuel for both the rural and urban populations has been the firewood followed by soft coke, kerosene oil, husk, hard coke and electricity in order of consumption level. The average commercial fuel consumption is found to be about 13.7 kg of coal equivalent at the present level of consumption of which firewood contribution alone is about 8.5 kg of equivalent coal per person per month.

Non-Commercial Fuels

The non-commercial fuels have been considered as those which the consumer manages to get from his own resources like from his farm, orchards, households, livestock etc. These include firewood, husk, dung, softcoke and other such fuels.

The average consumption of non-commercial fuel in different fuel forms is given in Table 3.10. The non-commercial fuel consumption pattern for rural and urban populations respectively has similarly been given in Tables 3.11, 3.12. It may be noted from Table 3.10 that the major non-commercial fuels in use are the firewood and dung and that the use of other forms of non-commercial fuels is almost negligible on the average. Comparison of the commercial and non-commercial fuel consumption data in Tables 3.7 and 3.12 suggests certain general trends as given below:

1. Firewood consumption dominates both the commercial and non-commercial fuel consumptions.
2. The average non-commercial firewood consumption is almost

double that of commercial firewood consumption. However, the commercial firewood in urban areas is almost five times the commercial firewood consumption in rural areas.

3. The non-commercial firewood consumption in rural areas is more than 3 times the non-commercial firewood consumption in urban areas.
4. The average total (commercial + non-commercial fuel consumption/person/month in urban population is somehow lower (29.4 kg of coal equivalent) in comparison to total (commercial + non-commercial) fuel consumption/person/month for rural population (40.53 kg of coal equivalent). This is for the fact that the rural sector consumes more energy both because of its inefficient utilization, the eating and cooking habits (3-4 times a day) and as well as to prepare the live-stock feed. It is also evident that whereas the rural sector is capable of producing 18 per cent of its energy needs, the urban sector is capable of producing only 30 per cent for its energy needs.

The correlation coefficients of the consumption patterns although not very strong with respect to the education and economy of the population, it is clear from Table 3.10 that correlation is quite strong with respect to education and economy in relation to use of firewood, kerosene oil and electricity.

The districtwise fuel needs of the State (both commercial and non-commercial in coal equivalent have been summarized in the Table 3.3. Therefore the fuel consumption for the J&K State at the present consumption level = 2,700 MWh obtained by converting the coal equivalents into electricity units. However, if the approximated of about 7 per cent (considering the optimal level forecast suggested by the working group on energy policy) both because of the rise in population and demand because of improved economics are considered over the successive plan periods, the following consumption levels are predicted.

TABLE 3.3: Fuel Consumption Patterns of the J & K State

District	Commer./month Coal Eq. Kg x 10		Commer./month Coal Eq. Kg x 10		Total/month Coal Eq. Kg x 10		Grand total/ month Coal Eq. Kg x 10
	Urban	Rural	Urban	Rural	Urban	Rural	
Anantnag	5.98	3.07	3.21	10.9	9.19	13.97	23.16
Pulwama	3.82	1.81	2.05	6.45	5.87	8.26	14.13
Srinagar	7.07	3:11	3.80	11.06	10.87	14.17	25.04
Badgam	3.26	1.06	1.75	5.7	4.86	7.03	12.16
Baramulla	1.33	5.36	0.07	19.25	2.03	24.6	26.63
Kupwara	0.51	2.67	0.28	9.7	0.79	12.37	13.16
Kargil	0.08	0.53	0.04	1.89	0.12	2.42	2.54
Leh	0.15	0.54	0.07	1.92	0.22	2.46	2.68
Doda	0.65	3.47	0.35	9.48	1.00	12.95	13.95
Udhampur	0.80	3.65	0.43	12.96	1.23	16.61	17.84
Kathua	0.50	3.02	0.27	10.7	0.77	13.72	14.49
Jammu	3.8	5.52	2.05	19.6	5.85	25.12	30.97
Rajouri	0.24	2.05	0.13	9.01	0.37	11.60	11.97
Poonch	0.30	1.83	0.16	6.51	0.46	8.34	8.80
J&K	28.49	38.67	15.29	135.13	43.78	173.80	217.58

Year	Per capita Eq. Coal Energy Consumption/month	Electricity Per capita/month KWh
1985-90	43.09	387.8
1990-95	47.00	423.0
1995-2000	50.29	452.6

TABLE 3.4: **Energy for Household Operation Man-Hrs/Month**

Operation	Average/ person	Eq. Coal kgx10^3	Correlations Coeff.	
			Education	Economy/ person
Preparation of Meals	20.02	71.9	–.02	.04
Meal Carrying to Farm	1.04	3.70	–.14	.07
Fetching of Water	3.00	10.68	–.15	–.02
Rice Cleaning	1.11	3.95	–.01	.06
Making Flour	0.95	3.38	–.06	.08
Vegetable Peeling	1.25	4.45	–.03	.04
Milk Processing	1.09	3.88	–.07	.09
Washing Clothes	1.42	6.05	.13	.05
Ironing	0.17	0.60	.22	.03
Mud Washing	2.96	10.53	–.00	.06
Drying of Grains	0.21	0.74	–.06	.12
Others (Miscellaneous)	8.18	29.12	.09	–.09
Total	147.98 kg. coal equivalent/person/month			

TABLE 3.5: **Energy for Household Operations Man-Hrs/Month (Urban Population)**

Operation	Average/ person	Correlation Coeff.	
		Education	Economy/ person
Preparation of Meals	19.41	–.04	.01
Meal Carrying to Farm	–0.16	–.09	.00
Fetching of Water	0.83	–.10	–.02
Rice Cleaning	1.26	–.00	–.02
Making Flour	0.58	.94	.07
Vegetable Peeling	0.83	.05	.02
Milk Processing	0.29	–.06	.07
Washing Clothes	2.17	.09	–.01
Ironing	0.38	.20	.08
Mud Washing	3.17	–.04	–.03
Drying of Grains	0.06	–.01	.02
Others (Misc.)	5.89	.17	.00

Total = 35.03 man hrs/month/person

TABLE 3.6: **Energy for Household Operation Man-Hrs/Month**

Operation	Average/ person	Correlation Coeff. Education	Correlation Coeff. Economy/ person
Preparation of Meals	20.21	–.01	.05
Meal Carrying to Farm	1.31	–.08	.09
Fetching of Water	3.68	–.07	–.02
Rice Cleaning	1.06	–.04	.07
Making Flour	1.06	–.05	.09
Vegetable Peeling	1.38	–.01	.05
Milk Processing	1.34	–.00	.10
Washing Clothes	1.19	.05	.06
Ironing	0.11	.16	.01
Mud Washing	2.90	–.00	.07
Drying of Grains	0.26	–.03	.15
Others (Misc.)	8.89	.12	–.10

Total = 43.39 man-hrs/month/person

TABLE 3.7: **Fuel Consumption (Commercial/Month)**

Fuel	Average/ Person	Coal eq. kg	Correlation Coeff. Education	Correlation Coeff. Economy/ person
Firewood (kg)	14.02	8.51	.13	.02
Husk (kg)	2.0	0.92	.01	–.02
Dung (kg)	0.2	0.12	.05	.02
Kerosene oil (liters)	1.24	1.48	.27	.09
Diesel (liters)	0.03	0.03	–.01	–.02
Electricity (Units)	2.39	0.26	.15	.07
Cooking gas (Cylnds)	0.02	0.003	.05	.04
Hard Coke (kg)	0.28	0.28	.02	.02
Briquettes (kg)	0.005	0.002	.04	–.01
Match Boxes (No.)	1.44	—	.08	.04
Soft Coke (kg)	5.27	2.26	.02	.02
Petrol (liters)	0.03	0.035	–.01	.01
Others (Misc.) (kg)	0.10	0.06		–.01

Total 13.68 kg. Coal equivalent/month/person

TABLE 3.8: **Fuel Consumption (Commercial Month) Urban Population**

Fuel	Average/ Person	Coal eq. kg	Correlation Coeff. Education	Correlation Coeff. Economy/ person
Firewood (kg)	20.08	12.19	—.02	.03
Husk (kg)	2.95	1.36	—.13	—.05
Dung (kg)	0.35	0.20	.11	.03
Kerosene oil (liters)	1.71	2.05	.09	.23
Diesel (liters)	0.01	0.01	.11	.00
Electricity (units)	2.52	0.28	.09	—.02
Cooking gas (cylnd)	0.03	0.03	.13	.01
Hard Coke (kg)	0.21	0.21	.06	.15
Soft Coke (kg)	6.27	2.70	—.03	.08
Briquettes (kg)	0.01	0.01	.04	—.02
Match Boxes (No.)	1.56	—	.02	.09
Petrol (liters)	0.01	0.01	.13	—.04
Others (Miscellaneous)	0.13	0.08	—.10	—.02

Total = 19.12 kg Coal equivalent/person/month

TABLE 3.9: **Fuel Consumption (Commercial)/Month (Rural Population)**

Fuel	Average/ Person	Eq. Coal kg.	Correlation Coeff. Education	Correlation Coeff. Economy/ person
Firewood (kg)	8.03	4.87	.10	.03
Husk (kg)	0.98	0.45	—.01	—.00
Dung (kg)	0.11	0.06	—.01	.01
Kerosene oil (liters)	0.78	0.94	.17	.03
Diesel (liters)	0.06	0.07	—.00	—.02
Electricity (unit)	2.26	0.26	.13	.08
Cooking gas (cylnd)	0.01	0.01	.01	.03
Hard Coke (kg)	0.35	0.35	—.01	—.01
Soft Coke (kg)	4.26	1.83	—.03	.02
Briquettes (kg)	0.00	0.00	—.01	.01
Match Boxes (No.)	1.32	—	.08	.04
Petrol (liters)	0.05	0.05	—.01	.01
Others (Misc.) (kg)	0.03	0.02	—.02	.01

Total = 8.91 kg Coal equivalent/person/month

TABLE 3.10: **Fuel Consumption (Non-Commercial)/(Month)**

Fuel	Average/ Person	Coal Eq. kg	Correlation Coeff. Education	Correlation Coeff. Economy/ person
Fire-wood (kg)	28.00	17.36	–.15	.04
Husk (kg)	0.01	0.004	–.02	–.01
Dung (kg)	15.84	9.23	–.04	.00
Soft Coke (kg)	0.02	0.01	–.01	–.00
Others (Misc.) (kg)	0.03	0.02	–.02	–.00

Total = 26.6 kg of coal/person/month

TABLE 3.11: **Fuel Consumption (Non-Commercial/Month) (Urban Population)**

Fuel	Average/ Person	Coal eq. kg	Correlation Coeff. Education	Correlation Coeff. Economy/ person
Firewood (kg)	10.56	6.40	–.06	–.03
Husk (kg)	0.02	0.01	.02	.03
Dung (kg)	6.86	3.87	.09	.07
Soft Coke (kg)	0.00	0.00	1	1
Others (kg)	0.00	0.00	1	1

Total = 10.28 kg coal equivalent/person/month

TABLE 3.12: **Fuel Consumption (Non-Commercial)/Month**

Fuel	Average/ Person	Coal eq. kg	Correlation Coeff. Education	Correlation Coeff. Economy/ person
Firewood (kg)	34.26	20.8	–.07	.07
Husk (kg)	0.10	0.004	–.02	–.01
Dung (kg)	18.62	10.80	.04	.01
Soft Coke (kg)	0.01	0.004	.00	–.01
Others (Miscellaneous) (kg)	0.03	0.02	–.01	–.00

Total = 31.62 kg of Coal equivalent./person/month

FOUR

SPATIAL AND TEMPORAL ANALYSIS OF ENERGY DEMAND IN JAMMU AND KASHMIR

An Inquiry into Consumption, Requirements and Predictions

NISAR ALI

INTRODUCTION

Food, work and productivity have been identified as the basic objectives of the Seventh Five Year Plan. Energy and its efficient use is one of the main determinants of production. Energy problems of the present dimensions emerged in the late seventies when the world petroleum market resorted to a drastic and dramatic price hike and consequently energy supply and demand became a matter of serious concern particularly in developing nations. Thus, various models under system analytic approach concerning the demand, supply and predictions for a finite time horizon have been attempted over the period of time. The attempts are being made to find out most viable, both from technical and economic points of views, alternative sources of energy in the contemporary world.

The main sources of energy can be broadly divided into two classes: commercial and non-commercial. The former includes coal, crude oil, natural gas, nuclear energy, hydro and thermal electricity, while the latter includes traditional fuelwood, animal waste, animal power, bio-gas and solar applications.

The problem of energy in India has become acute since the world energy crisis of the early seventies. On analysing the future demand and supply India does not appear to be richly endowed in terms of hydrocarbons. The present annual growth of national income and population necessitates search for more and more energy resources and alternatives. The energy consumptions and requirements vary from region to region depending on the agro-climatic conditions and altitude considerations. In the State of Jammu and Kashmir the domestic sector is consuming a significant proportion of non-commercial energy particularly fuelwood, thus exerting more pressure on the valuable forest wealth. It becomes an imperative for planners to find out some trade off between fuelwood and other alternative source of energy. Since there is great potential of hydro-power generation, about 10,000 MWs–, serious effort is required to utilize this potential. This will not only raise the State's income but will also maintain the eco-system of the State.

Brief Review of Studies

The energy problems of the developing countries are more acute than of the developed ones. The developing countries which constitute the world's 71 per cent[1] population consume only half as much commercial energy annually as the USA which constitutes 5.5 per cent of the world's population. The disparity is more striking when the same is evaluated on per capita basis. The per capita energy consumption in the developing countries was 0.5 tce in 1975.[2] The disparities among the developing nations are still more striking.

India meets most of its energy requirements from domestic sources. Three-fourths of the crude oil is imported, yet the same constitutes 20-25 per cent[3] of the total consumption of commercial energy. The large and easily accessible coal deposits in the country, though of low quality, do not fulfil the requirements and to some extent its uneven distribution is off-set by the availability of hydro-electric power which accounts for 40 per cent of the total power generation as against 5 per cent of the nuclear power. At present one-sixth of the hydro-potential is utilized at the national level.

Yet, a substantial proportion of the country's consumption requirements of energy comes from non-commercial sources, fuelwood being the most important. The non-commercial energy accounts for a little over 50 per cent[4] of the total energy consumption out of which fuelwood accounts for 65 per cent, cowdung 15 per cent and crop wastes 20 per cent. Since 70 to 80 per cent of the population lives in rural areas, the non-commercial energy consumption constitutes 91.1

per cent[5] of the total rural non-commercial energy consumption, while the commercial energy consumption constitutes 66 per cent of the total domestic energy consumption in urban India.[6]

The relative steady increase of commercial energy has been more than non-commercial energy over the period. In 1953-54 it was 32 per cent of the total energy which increased to 55 per cent in 1976-77.[7] In the case of non-commercial energy the percentage share has declined from 68 to 45 during the said period. The analysis of data of the mid-sixtees shows that income elasticity of demand for domestic consumption of energy is very high, that is, 2.12 which along with the population growth can still push up the demand[8] for energy in the foreseeable future. With the average growth of GDP at 3.55 per cent per annum between 1960-61 and 1975-76, the commercial energy consumption registered an annual growth of 6.35 per cent giving an income elasticity of demand 1.8. Among its main sources power registered a high growth rate of 9.5 per cent followed by oil products, 6.6 per cent and coal by 3.8 per cent per annum during the said period. The per capita power consumption at all-India level stands at 131 Kwhs as compared to 51 Kwhs in Himachal Pradesh, 211 Kwhs in Haryana, 303 Kwhs in Punjab and 581 Kwhs in Jammu and Kashmir. Looking at the energy consumption pattern the increase in the industrial sector has been twice that of the household sector from 1953-54 to 1975-76 in terms of coal replacement ratio.

Objectives

In the light of the tradition-oriented consumption pattern of energy on the one hand and high potential of hydro-power generation on the other, the present paper intends to make a study of energy problems in the State with the following objectives:

1. To identify commercial and non-commercial sources of energy.
2. To identify the hydro-power potential and its availability.
3. To study the growth pattern of power.
4. To study the spatial distribution of consumption pattern of energy in Kashmir.
5. To make predictions of power requirements and other sources of energy in the State (Kashmir).

Methods of Analysis

The present study is not based on the data generated by the study itself. It uses both time series and cross-section observations available from secondary sources. The findings of the study are based on these two types of data.

Time Series Observations

Power

The time series data on generation, installed capacity, imports and availability of power are provided by the State Government and is published.

Fuelwood

The data on fuelwood supply and distribution are provided by the Forest Department of the State. Although the data is available from 1950-51, it pertains to urban Srinagar only.

Kerosene Oil

There is no time series data available on this consumption. However, the Department of Food and Supplies provides the data on kerosene oil distributed by various depots in urban Srinagar for a few years. The same is used with the restricted assumptions in the present study.

Cooking Gas

The data on cooking gas cylinders and refills distributed to the various depots in Srinagar in aggregate is provided by Indian Oil/Bharat Petroleum, Srinagar Branch. However, the data is available only for a few years.

Cross-Section Observations

There is no study available on the energy consumption in Jammu and Kashmir State based on survey data with reference to any period. The two sources of crude data, however, can be used as cross-section data which are collected by two different agencies. These are given below:

Power Data

The energy accounts provided by the Power Development Department can be used as cross-section observations. The accounts have been prepared by the said department for Jammu and Kashmir Circles independently.

CORD Data

The present study has made use of raw data collected by the Centre of Research for Development, University of Kashmir, on renewable energy resources. Since the report based on the said data is not available no sample design and methodology could be spelt out in the present study while making use of the data. However, it is presumed that the collected data is based on an appropriate and scientific methodology while the study was carried out.

Wood Balance Pre-Testing Study

A wood balance study conducted by the Directorate of Economics on behalf of the World Bank's Social Forestry Project, Jammu and Kashmir Government, has been conducted very recently. The study is based on a sample of 416 households drawn from Dry, Temperate, West Zone and Sub-Tropical Zone of Jammu and Kashmir State. The study has been carried out for pre-testing for drawing an appropriate sample and the report is yet to be published. Therefore, the reference is used in the present study or paper only to make a comparison with CORD data and is not quoted here as a published document.

NSS Data

We have also used NSS data on Fuel and Lighting to get the elasticity parameters for the year 1968-69 and 1973-74.

Reference Period

The energy accounts were prepared for the year 1983-84, the CORD survey was conducted in 1981-82 and the wood balance study was carried out from August 1983 to July 1984.

Data Limitations

Energy Accounts

While the energy accounts provide details of energy consumption and availability of power, the energy losses are arrived at by subtracting energy sold from energy pumped in at a given K.V. However, the technical losses have not been segregated from consumption losses which probably are of significant proportion. Therefore, it is difficult to ascertain an actual household consumption loss of power on account of pilferage.

CORD Data

Since the report on the said data has not probably been prepared till date, the underlying methodology of data collection is unknown. The sampling error is also not known. Thus, the reliability of the data could not be ascertained.

Wood Balance Data

The study aims at pre-testing for sample survey and the study is yet to be published. No logical inference can be drawn from a sample which is less than 0.04 per cent.

Assumptions

In order to carry out the analysis the following assumptions are made:

1. The decennial growth rate of population between 1971 and 1981 will hold good for the projected period.
2. The average size of household for 1971 will hold good for 1981.
3. The consumption behaviour of households for the projected period will remain the same.
4. The rural-urban population migration will remain constant.
5. The transmission and distribution losses and auxiliary consumption of power will remain constant over time.

6. The CORD data follows a scientific methodology of data collection and is subject to minimum sampling error.
7. The household consumption behaviour based on small samples will not be significantly different from a behaviour based on large samples.
8. The growth rate of per capita State income, that is, 2.27 per cent per annum from 1973-74 to 1982-83 will remain the same during the projected period.
9. We also assume that the income elasticity parameter both rural and urban for energy consumption will remain constant over the projected period.

The Model

In order to find out net and gross availability of power and energy for Jammu and Kashmir independently the following equations are used:

(A)

$$Eg = Ne + Ie \quad \ldots (1)$$

$$Le = Re + De \quad \ldots (2)$$

$$Nae = Eg - Le$$

$$= Ne + Ie - [Re + De] \quad \ldots (3)$$

$$Nek = Ek + Iek - Xek - Rek \quad \ldots (4)$$

$$= Ek + Iek [Xek + Rek] \quad \ldots (5)$$

$$Nej = Nae - Nek \quad \ldots (6)$$

$$= Ne + Ie - [Re + De] - Ek + Iek - [Xek + Rek] \quad \ldots (7)$$

where

Eg = Gross availability of energy in Jammu and Kashmir State
Le = Total leakage
Re = Inter-regional losses
De = Energy despatched to Salal
Nae = Net availability of energy in J&K
Ne = Net generation of energy in the State
Ie = Imports from neighbouring States
Nek = Net availability of energy in Kashmir
Ek = Net self generation of power
Iek = Imports from Udumpur to Bemina as recorded at Udumpur

Xek = Exports to Udumpur from Bemina as recorded at Bemina
Rek = Inter-regional losses of Udumpur-Bemina line
Nej = Net availability of energy in Jammu

(B) $Ec = Ep - Xu - Lo$
$= Ep - [Xu + Lo]$ .... (8)

where

Ep = Energy consumption/sold
Xu = Auxiliary consumption
Lo = Losses

In order to find out the growth of power the exponential function of the following variety has been fitted to the time series data on installed capacity of power, power generation, power imported and power availability in Kws. The function takes the following form:

$$Y = a\,b^{X} \quad \text{.... (9)}$$
$$\text{or } \log Y = \log a + X \log b + u \quad \text{.... (10)}$$

where
Y = Time series dependent variable
X = Time, 1950-51 to 1983-84

Energy forecasts can be made by using two methods. First, to find the per capita power consumption availability and multiply with the projected population for the projected period under a given set of assumptions. Secondly, with the per capita availability/consumption of energy, forecasts can be made by taking income elasticity parameter as exponent and multiplying it with projected population.

The long-term forecasts in the present study have been made on the basis of Engel elasticities computed from NSS data. The Engel function of the following form have been used for predictions.

$$\log Y = \log \alpha + \beta \log X + u$$

where
Y = per capita expenditure on fuel and lighting per month

X = Aggregate expenditure per month per capita
α = Intercept
β = Elasticity coefficient

Growth, Problems and Potential of Power

The State has an acute shortage of power. But extensive availability of power is necessary for self-sustained growth of the State economy particularly for expanding the industrial base and extension of irrigation to the areas where irrigation through gravity canals is not possible. However, about 10,000 MWs of hydro-power potential has been identified in the State out of which only 2 per cent has been harnessed.

Power Potential

Most of the power in the State is generated from hydel projects. The power generating capacity has increased from 11 MWs in 1955-56 to about 209 MWs in 1982-83. The installed capacity of the existing power projects is given below:

		Installed Capacity (1982-83MWs)
1.	Ganderbal (Hydro)	15.00
2.	Upper Sindh (Hydro)	22.60
3.	Mohra (Hydro)	9.00
4.	Lower Jehlum (Hydro)	105.00
5.	Chinani (Hydro)	23.60
6.	Kalakotes (Thermal)	22.50
7.	Others (Small Units)	11.23
	Total	208.93

Out of the total power potential identified, a potential of 179 MWs stands commissioned in the State, while the several projects with the potential of 660 MWs are under construction and a good number are under investigation. The State being rich in water resources the hydro-power potential utilization over time may lead the State economy to take off for self-sustained growth in the near future.

Growth of Power

We have found out the annual rate of increase in power generation on a five-yearly basis. According to it the installed capacity witnessed an increase of 15.38 per cent per annum during the First Plan period; the power generation decreased marginally consequently causing a significant rise in power imports (Table 4.1) that is, 72.23 per cent per annum. This may be attributed to low priority given to the power sector in the First Five Year Plan. It seems that the installed capacity during the last few years is showing a decelerating trend while the power generation shows a marginal increase per annum, that is, 5.27 per cent.

TABLE 4.1: **Rate of Annual Increase in Installed Capacity, Power Generation Imports and Power Availability on Five-Yearly basis**

(in percentage)

Year	Installed Capacity (MWs)	Power Generation (KWs)	Imports (KWs)	Availability (KWs)
1950-51 to 1955-56	15.38	–0.21	72.23	3.78
1955-56 to 1960-61	1.45	14.16	27.50	16.98
1960-61 to 1965-66	24.77	27.92	30.97	28.75
1965-66 to 1970-71	9.42	14.50	5.51	12.07
1970-71 to 1975-76	31.35	16.19	11.42	15.12
1975-76 to 1980-81	20.02	30.82	52.38	35.50
1980-81 to 1983-84	–0.02	5.27	16.17	7.98

In order to find out the rate of growth of power during the reference period we have fitted a growth model of exponential variety to the time series date of the following form:

$$Y = ab^{X}$$

where,

Y = Installed capacity, Power Generation (Kws), Imports (Kws) and Power Availability(Kws)

X = Time, 1950-51 to 1983-84

a & b = Constants

We have arrived at the following estimated equations:

$$\log Y = \log \alpha + X \log \beta + u$$

Inst. Cap

$$\log Y = 4.02 + 0.0672X \quad R^2 = 0.92$$
$$(0.0044) \quad SE = 0.0044 \quad Sig = 0.001 \text{ level}$$

Power Gen

$$\log Y = 2.53 + 0.0698X \quad R^2 = 0.94 \quad Sig = 0.001 \text{ level}$$
$$SE = 0.0043$$

Power Impt.

$$\log Y = 1.92 + 0.0822 \quad R^2 = 0.80 \quad Sig = 0.001 \text{ level}$$
$$SE = 0.0092$$

From the above equations the annual exponential growth is arrived at by (b – 1)X100 which indicates that the installed capacity, power generated and imports have registered annual rates of growth of 16.74, 17.45 and 20.84 per cent respectively during the reference period.

The growth analysis further reveals that the State has been continuously depending on power to maintain the existing rate of consumption and availability.

Power Availability

Although the availability of power both from imports and self-generation has registered a significant increase over the period of time, yet out of the existing generating capacity of 209 MWs, the effective utilization of capacity is still less, that is, 168 MWs and 12 MWs during the summer and winter months respectively. Since most of the power generation is hydro by nature, the scarcity of water reduces the self-generation during the winter months, with the result the power deficits continue to increase unabated. During 1980-81 the power shortage was 24 MWs and 91 MWs in the summer and winter months respectively and the deficits increased to 103 MWs and 187 MWs during the summer and winter seasons respectively in 1983-84. As regards the energy the deficit on this count has gone up manifold from 32 million units in 1980-81 to 380 million units in 1983-84. Consequently the per capita availability of energy in the State is very low when compared with many other States. The per capita availability of energy increased from 60 Kwhs in 1973-74 to 193 Kwhs in 1984-85. During the last two years the per capita availability has declined from 204 Kwhs to 193 Kwhs, according to official estimates.[13]

The availability of power depends, among other things, on inter-regional transmission, distribution losses and auxiliary consumption. Our trend result shows that the power availability has registered a growth rate of 18.02 per cent per annum during 1950-51 to 1983-84 which is given by the following estimated equation:

$$\log Y = 2.63 + 0.0719X \qquad R^2 = 0.92 \qquad SE = 0.0040$$

Sig level 0.001

It appears that transmission/distribution losses at present are very high thus reducing the power/energy availability (Table 4.2). However, the State has been able to reduce total losses from 169 lakh units in 1982-83 to 91.87 lakh units.

Loss of Power

The power losses are due to two counts: first, Technical, i.e. Transmission and Distribution losses; second, Consumption, i.e. losses which are on account of pilferages by household consumers and are unaccounted consumption. The overall losses have been reduced from 51.28 per cent in 1979-80 to 37.04 per cent in 1982-83. The data reveals that the proportion of losses has throughout been higher in the Kashmir than in the Jammu region.

Power Consumption

Temporal Analysis

In the present paper the energy sold and energy consumed are used as synonyms. The energy consumption has registered a persistent increase from 1971-72 to 1983-84. The per capita consumption of energy has risen from 38.65 Kwhs in 1971-72 to 103.82 in 1983-84. The official estimates reveal that the domestic consumption of energy has declined over the period and energy consumption by the industrial sector also

TABLE 4.2: **Distribution of Energy region-wise in terms of Gross Self-Generation, Net Self-Generation, Imports, Inter-Regional Losses and Availability during 1982-83 and 1983-84. (in LUs)**

Region	Year	Gross Self-Generation	Net Self-Generation	Imports	Gross Availability	Inter-Regional Losses	Energy supplied to Salal	Energy available
Kashmir	1982-83	8232.93	8208.04	266.4	8474.44	10.98	-	7096.27
Jammu	-do-	1969.47	1057.02	2732.95	3789.97	158.25	-	3935.75
J&K	-do-	9301.4	9265.06	2999.35	12264.41	169.23	359.03	11032.02
Kashmir	1983-84	7972.06	7948.26	—	—	—	—	—
Jammu	-do-	905.28	892.45	—	—	—	—	—
J&K	-do-	8877.34	8840.71	4226.32	13067.02	91.87	424.45	12550.70

Note: Compiled from Energy Accounts for 1982-83 and 1983-84.
Source: Maintenance and R.E. Wing Jammu, PD.D Deptt., Oct, 1984.
LUs = Lakh units

has gone down. Due to the expansion of the tertiary sector the energy consumption must have gone up during the reference period. While in 1971-72 domestic energy consumption constituted 42.43 per cent of the total consumption, the same accounts for 27 per cent in 1983-84. The reduction in losses along with the fall in domestic consumption may lead one to conclude that unaccounted consumption might have gone up (Table 4.3).

Spatial Analysis

The domestic consumption of energy in the Kashmir region is twice that of the Jammu region, while there is not much difference between the energy consumption by the industrial sectors in these two regions. Even for the sectors like commerce and agriculture the difference is not so sharp between these two regions. In the Valley the domestic energy consumption in South Circle or District Anantnag is more than 52 per cent compared to District Srinagar which accounts for 29 per cent only (Table 4.4) and Baramulla 36.31 per cent.

TABLE 4.3: **Distribution of Sector-wise Energy Consumption from 1971-72 to 1983-84 in Jammu and Kashmir State**

(in lakh KWs)

Year	Domestic	Commercial	Industrial	Others	Total	Per capita
1971-72	757	135	633	259	1784	38.65
	(42.43)	(7.57)	(35.48)	(14.52)	(100)	
1972-73	366	177	715	818	2076	43.00
	(17.63)	(8.53)	(34.44)	(39.40)	(100)	
1973-74	888	195	728	349	2160	43.76
	(41.11)	(9.03)	(33.70)	(16.16)	(100)	
1983-84	1700.41	366.34	1689.47	2573.81	6330.03	103.82
	(26.86)	(5.79)	(26.69)	(40.66)	(100)	

Notes: Figures in parentheses are percentages.
Data for 1983-84 is in lakh units.

Source: Digest of Statistics, Directorate of Eco. & Stat. Deptt. of Planning and Development, J&K Govt.

The higher energy consumption in the industrial sector is found in urban regions like two main cities, Jammu and Srinagar. The per capita energy consumption stands at 115 units in Kashmir Province and 86

TABLE 4.4: **Sector-wise Distribution of Energy Consumption according to Region/Circle in Jammu and Kashmir State**

(in lakh units)

Region/ Circle	Domestic	Commercial	Industrial	State/ Central	Agriculture/ Irrigation	Bulk Supply and Army	Others	Total
Srinagar	739.26	206.02	789.28	140.30	57.83	433.38	198.92	2564.99
	(28.82)	(8.03)	(30.77)	(5.47)	(2.25)	(16.90)	(7.76)	(100)
Baramula	245.82	16.91	142.90	24.54	91.88	88.94	66.02	677.01
	(36.31)	(2.50)	(21.11)	(3.62)	(13.57)	(13.13)	(9.76)	(100)
Anantnag	299.31	19.64	116.92	24.54	34.02	20.57	55.46	570.46
	(52.47)	(3.44)	(20.50)	(4.30)	(5.96)	(3.61)	(9.72)	(100)
Kashmir Division	1284.39	242.57	1049.10	189.38	183.73	542.89	320.40	3812.46
	(33.69)	(6.36)	(27.52)	(4.97)	(4.82)	(14.24)	(8.40)	(100)
Jammu	267.99	83.13	438.15	129.95	114.73	233.41	378.11	1645.47
	(16.28)	(5.06)	(26.63)	(7.90)	(6.97)	(14.18)	(22.98)	(100)
Kathua	39.41	9.29	170.11	22.28	5.94	—	86.27	333.30
	(11.82)	(2.79)	(51.04)	(6.68)	(1.78)	(–)	(25.89)	(100)
Udumpur	52.99	20.97	20.09	35.11	5.96	142.19	67.40	344.71
	(15.37)	(6.09)	(5.83)	(10.18)	(1.73)	(41.25)	(19.55)	(100)
Poonch	17.73	1.92	1.53	5.42	—	—	10.89	37.49
	(47.29)	(5.12)	(4.08)	(14.46)	(–)	(–)	(29.05)	(100)
Rajouri	19.62	4.64	7.53	9.56	2.29	39.30	39.79	122.76
	(15.98)	(3.78)	(6.14)	(7.79)	(1.86)	(32.03)	(32.42)	(100)
Doda	18.38	3.82	2.96	8.05	—	—	0.64	33.85
	(54.30)	(11.27)	(8.75)	(23.78)	(–)	(–)	(1.90)	(100)
Jammu Region	416.12	123.77	640.37	210.37	128.92	414.93	583.10	2517.58
	(16.53)	(4.92)	(25.44)	(8.35)	(5.12)	(16.48)	(23.16)	(100)
Jammu and Kashmir	1700.51	366.34	1689.47	399.75	312.65	959.20	903.50	6330.04
	(26.86)	(5.79)	(26.69)	(6.31)	(4.94)	(15.15)	(14.26)	(100)

Note: Figures in parentheses are percentages.
Source: Compiled from Energy Accounts.

units in Jammu Division. The higher per capita consumption is found in two districts, Srinagar and Jammu, that is, 224.93 units and 85.99 units respectively. The energy consumption ratio between the two provinces works out to be 2:3. The energy loss is more pronounced in the Kashmir region than in Jammu region. About 195 units per capita is the energy loss in Kashmir and only 33 units per capita in the Jammu region. However, among the districts in Jammu province, the per capita loss is higher in District Jammu than in other districts. The energy loss and urbanization seem to be positively correlated (Table 4.5).

The data further shows that the loss of energy is positively correlated with energy pumped in. It seems strange that the population and energy consumption by some crucial sectors in the Kashmir region is not highly positively correlated. This may be due to a higher proportion of loss by the domestic sector. The Correlation is high between population and sectoral energy consumption in the Jammu region.

Demand Prediction

According to official estimates the demand for energy will increase from 667 million units in 1983-84 to 770 million units in 1984-85 during the summer months and from 803 million units to 920 million units during the winter months in the reference years. The energy availability will go up from 579 million units to 606 million units during the summer months and from 423 million units to 440 million units during the winter months between 1983-84 and 1984-85. These relativities show that the summer deficit will increase from 88 millions units to 164 million units during the summer months and 380 millions units to 480 million units during the winter months between 1983-84 and 1984-85.

However, on the basis of per capita consumption of energy in 1983-84 the demand for energy will increase by 4,122.72 lakh units in 1985 and 4,640 lakh units in 1990 for Kashmir and 2,699.03 lakh units in 1985 and 3,061 lakh units in 1990 for Jammu. It is estimated that the total demand for power consumption will increase by 6,821.75 lakh units in 1985 and 7,701 lakh units in 1990 for the whole State. These projections have been arrived at by using a simple method.

TABLE 4.5: **Distribution of Energy Consumption and net availability and Energy Loss in Aggregate and Per Capita according to Regions/Circles in Jammu and Kashmir, 1983-84**

(in Lakh Units)

Region/ Circle	Net avail-ability	Consump-tion/Sold	Total population	Percentage of		Density per sq. km	Net energy available		Energy consump-tion per capita	Energy loss per capita
				Available	Consump-tion		per worker	per capita		
Srinagar	4787.50	2564.99 (53.58)	1140361	47.37	40.74	378.48	1123.35	419.82	224.93	194.90
Anantnag	893.53	570.45 (63.84)	1158183	8.84	9.06	215.20	184.83	77.15	49.25	27.90
Baramulla	1098.80	677.01 (61.61)	1055762	10.87	10.75	1415.36	236.23	104.08	64.12	39.95
Kashmir	6779.83	3812.45 (56.23)	3314706	67.08	60.55	207.78	493.17	204.54	115.02	89.52
Jammu	2269.05	1645.47 (72.52)	997969	22.32	26.15	315.32	606.36	227.37	164.88	66.26
Rajouri	169.48	122.76 (72.44)	326193	1.68	1.95	121.67	124.55	51.95	37.63	18.26

TABLE 4.5 *Contd.*

Region/ Circle	Net avail- ability	Consump- tion/Sold	Total population	Percentage of Available	Percentage of Consump- tion	Density per sq. km	Net energy available per worker	Net energy available per capita	Energy consump- tion per capita	Energy loss per capita
Kathua	368.42	333.30 (90.46)	394509	3.64	5.29	148.82	229.97	93.39	84.48	12.16
Udumpur	426.83	344.71 (80.76)	484372	4.22	4.92	106.48	172.39	88.12	71.17	25.46
Poonch	48.26	37.49 (77.68)	238220	0.48	0.60	143.68	46.25	20.26	15.73	9.92
Doda	58.26	33.85 (58.10)	445891	0.57	0.54	38.14	31.60	13.07	7.60	8.36
Jammu Region	3340.30	2517.58 (75.68)	2887154	32.92	39.45	109.80	276.17	115.22	85.99	33.03
J&K State *	10120.13	6330.04 (62.64)	6201860	100	100	146.45	—	162.96	101.51	63.22

Notes: 1. Population is projected on 1971-1981 growth rate.
2. Figures in parentheses are percentages.

TABLE 4.5A: **Regionwise Coefficients of Correlation**

Region	Variables	Coefficient of Correlation
Kashmir	Energy Pumped-in/Loss	0.92
Jammu	Energy Pumped-in/Loss	0.86
J&K	Energy Pumped-in/Loss	0.89
Kashmir	Net Availability/Loss	0.92
Kashmir	Energy Sold/Loss	0.66
Jammu	Energy Sold/Loss	0.72
J&K	Energy Sold/Loss	0.65
Kashmir	Population/Domestic Consumption	0.45
Kashmir	Population/Commercial Consumption	0.24
Kashmir	Population/Industrial Consumption	0.32
Jammu	Population/Domestic Consumption	0.95
Jammu	Population/Commercial Consumption	0.91
Jammu	Population/Industrial Consumption	0.84

The per capita income during 1973-74 and 1983-84 has registered a growth rate of 2.27 per cent per annum. Assuming the income per capita grows at the same rate during the coming decade and further assuming that the estimated income elasticity parameter remains constant we can predict the energy consumption and requirements for 1985 and 1990 by the following method:

$$Q_t = Q_o \left(1 + \frac{r}{100}\right)\eta$$

where

Q_t = Per capita energy consumption in year t

Q_o = Per capita energy consumption in base year 1983-84

r = Percentage increase in per capita State income at constant prices.

η = Income elasticity of demand for energy based on 1973-74 NSS data.

According to the above method the consumption demand for power (energy) for 1985 and 1990 will be as under (Table 4.6 B):

TABLE 4.6: **Region and Circle-wise Distribution of Energy pumped in, Energy Loss, Energy Sold with percentages and the Sale Rate**

in lakh units

Region/Circle	Energy pumped in	Total energy loss	Percentage of loss	Energy sold	Percentage sold	Sale rate per unit (Rs.)	Revenue realised in Rs. lakhs
Srinagar	4876.3	2222.51	45.58	2564.99	52.60	0.27	707.39
Baramulla	1146.17	421.79	36.80	677.01	59.06	0.28	139.77
Anantnag	906.01	323.09	35.66	570.45	62.96	0.24	192.63
Kashmir Region	6928.48	2967.39	42.83	3812.45	55.03	0.27	1038.09
Jammu	2306.75	661.33	28.67	1645.47	71.33	0.32	531.31
Kathua	381.28	47.98	12.58	333.30	87.41	0.28	93.24
Poonch	61.12	23.63	38.67	37.49	61.33	0.42	15.83
Rajouri	182.34	59.57	32.67	122.76	67.32	0.35	42.78
Udumpur	478.03	123.32	25.80	344.71	72.11	0.80	276.27
Doda	71.56	37.71	52.69	33.85	47.30	0.42	14.21
Jammu Region	3481.07	953.54	27.39	2517.58	72.32	0.33	812.96
Jammu and Kashmir State	10409.55	3920.93	37.67	6330.04	60.80	0.30	1851.05

Note: Compiled from Energy Accounts. PDD, J&K Govt., 1984.

TABLE 4.6 A: The various sources of energy supplied in Srinagar from 1982 to 1984 are

Year	Qty of Fuel-wood disposed '000 Qts	Cooking Gas	K. Oil sold '000 Lits	Domestic consumption of Power (lakh units)	Total Energy Consumption		
					Total	P.m	P.a (mtce)
1982	411.14	—	22967	536.38	283322	0.40	4.8
1983	406.90	373.07	27126	657.3	266898	0.37	4.44
1984	392.28	909.62	25921.5	587.96	333881	0.43	5.16

TABLE 4.6 B: **Region-wise consumption demand for Power Energy**

Year	Region	Consumption Demand (lakh units)
1985-86	Kashmir	4161.43
	Jammu	2724.45
	J&K	6888.51
1990-91	Kashmir	4684.41
	Jammu	3089.56
	J&K	7781.02

Since the Engel parameter pertains to 1973-74, we can compound the predicted demand by taking time as an overall exponent so that the underestimation of the projected demand can be overcome. However, there is a possibility of over-estimation of the demand projection if t is taken into account.

The formula earlier mentioned can be modified as under:

$$Q_t = Q_o \left[\left(1 + \frac{r}{100}\right)\right]^t$$

where t stands for number of years between base year and projected year. The projected demand for energy is mentioned below:

Year	Region	Consumption Demand (lakh units)
1990	Kashmir	4907.93
1990	Jammu	3236.92
1990	Jammu and Kashmir	8152.38

The computed income elasticity of demand has also been used to predict fuelwood consumption and energy demand based on time series data. The fuelwood disposed of in urban Srinagar in 1984 was 3.92 lakh Qts and the kerosene oil sold was 25,921,000 Lits during the same year. We have converted these urban sources of energy into metric tonnes of coal equivalent and we find that per capita energy consumption has increased during 1981 to 1984 (Table 4.6 A).

The demand for fuelwood will increase to 513,000 qts and 611,000 qts in 1990 and 1995 respectively (in relation to 1985 demand). The predicted demand for energy is also mentioned below:

Year	Fuelwood demand '000 qts.	Total Energy Demand ('000 mtce)
1985	415.45	357.51
1990	513.25	441.63
1995	611.05	525.77

Cross-Section Analysis

The energy consumption sources in Kashmir are by and large traditional. Among the various sources of energy, the traditional fuelwood accounts for 46.32 per cent (CORD data) and 74.20 per cent (W.B. Study) followed by cowdung and coal. The pattern of energy consumption varies from place to place. While the CORD data shows no difference in the proportion of fuelwood consumption between urban and rural sector, the Wood Balance Study shows a considerable difference (Table 4.7).

The proportion of kerosene oil consumption in urban areas is higher than in rural areas and the reverse is shown by the data in the case of cowdung. Both the cross-section data show a same trend in respect of the said two domestic sources of energy. On an average the CORD study shows a higher per capita consumption of fuelwood than W.B. study. The proportion of fuelwood consumption in rural areas is 46 per cent but it is 78 per cent according to by W.B. study. While the CORD study shows 9.13 per cent consumption of kerosene oil in rural Kashmir, it is 2.1 according to W.B. study (Table 4.8 and Table 4.9).

Income Size and Energy Consumption

Energy consumption can broadly be said to be related to the level of income, family size, altitude and season. In the present analysis the aggregate income corresponding to each income size class is not provided in the W.B. study. Therefore, one cannot arrive at the income elasticity parameter on the data of W.B. study. However, it looks from the data that with the increase in the size of income the consumption of energy (in mtce) falls up to a certain level and thereafter it steeply rises as shown in Fig. 4.1.

TABLE 4.7: **Percentage Distribution of Energy Consumption by Household Sources of Energy in Kashmir according to Rural and Urban Sector.**

(in CRR)

Sector	Study	Fuelwood	Cowdung	Kerosene Oil	Electricity	Coal/Coke	Others	Total
R	CORD	46.43	30.65	9.13	2.44	11.35	—	100
	W.B.	78.2	16.5	2.1	—	neg	3.2	100
U	CORD	45.94	11.59	24.59	4.02	13.86	—	100
	W.B.	60.20	5.20	14.70	—	4.7	15.20	100
T	CORD	46.32	26.55	12.46	2.78	11.89	—	100
	W.B.	74.02	13.8	5.4	—	0.10	6.50	100

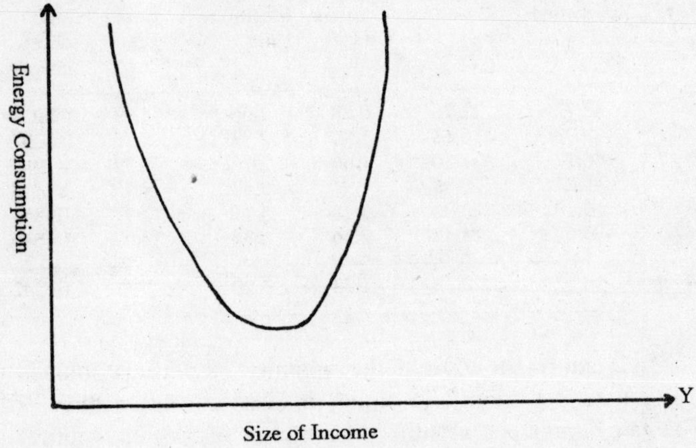

Fig. 4.1: U curve.

Thus income size and energy consumption behave according to U Curve. The U hypothesis also holds for cowdung. As the income tends to increase the traditional fuelwood is substituted by alternative sources of energy like kerosene oil, cooking gas, etc. The consumption of fuelwood increases as the income increases after certain income size. The fuel probably becomes cheap due to scale economies when consumed in large quantities with the result that high income strata consume more of fuelwood than other sources. However, one cannot confidently advocate this as income effect unless one knows the size elasticity. The W.B. study shows a pattern of consumption according to said hypothesis (Tables 4.8 and 4.9).

TABLE 4.8: **W.B. Study: Distribution of per capita consumption of energy per month according to size of income (1983-84)**

Income size (Rs.)	Fuelwood (kgs)	Cowdung (kgs)	K.Oil (Lit)	Coal (kgs)	Cooking Gas (kgs)	Total Eng Cons. (qts tce)
Less than 500	42.34	8.58	0.14	—	—	0.5425
500-1000	37.08	10.84	0.30	—	—	0.5398
1000-1500	26.67	7.50	0.89	—	0.02	0.4415
1500-2000	27.50	6.67	1.20	—	0.43	0.4967
2000-3000	25.00	4.17	1.55	0.8	0.71	0.4989
3000 & above	28.67	3.67	2.20	—	1.07	0.5989
All Classes	31.67	8.34	0.73	—	0.17	0.5007

TABLE 4.9: **CORD Study: Per Capita Consumption per month according to Income Size**

Income Size (Rs)	Fuelwood & Husk kgs	Cowdung kgs	Kerosene Oil Lits	Electricity Units	Coal kgs	Total energy (mtce)
/500	43.72	22.78	0.80	2.06	4.09	0.9054
500 -1000	39.92	16.13	1.32	2.73	5.75	0.8397
1000-1500	60.17	17.30	1.66	2.75	13.34	1.1965
1500-2000	46.03	20.88	1.94	3.42	9.95	1.7716
2000-2500	63.18	20.00	3.28	4.11	11.37	1.3841
2500 & above	84.02	14.42	2.36	4.34	14.05	1.4646
All Classes	56.07	20.35	1.73	3.20	9.37	1.1542

The CORD study shows that the quantity of various sources of energy per capita per month is higher in absolute terms than W.B. study. The per capita per month consumption of various sources of energy works out to be 1.15 mtce or 115 qts (CORD) as against 1/2 qts tce according to W.B. study. Thus there appears a significant difference in consumption of energy in these two studies. While the former shows decline in the fuelwood consumption with the increase in income the latter leads one to reverse conclusions (Tables 4.8 & 4.9). Since the W.B. study satisfies the technical characteristics excepting the size of sample and CORD study does not provide the technical characteristics with regard to the sample, one has reason to have doubts on the latter study. The two reference data should not show so obvious a difference.

Regional Profile

The fuel consumption stand at 61 kgs per capita per month in rural Kashmir according to CORD study and the same is put at 37 kgs by W.B study. For urban Kashmir it works out to be 48 kgs, according to the former study and 24 kgs per month according to latter. Among the districts the fuelwood consumption appears to be higher in Anantnag followed by Srinagar and Baramulla (CORD). The fuelwood consumption per capita is higher in district Baramulla according to W.B. study. The former shows 1.77 lits of kerosene oil consumption per capita per month while the latter indicates 1/2 lit per capita per month (Tables 4.10 & 4.11). On an average the household energy

TABLE 4.10: **Per Capita and Per Household Consumption of Energy by Source Per Month and Per Year according to Districts and Rural-Urban Sectors in Kashmir (1981-82)**

District	Region	Fuelwood (qts)	Cowdung (qts)	Kerosene Oil (lits)	Coal (qts)	Electricity (units)	Total Energy (mtce)	Total Energy (per annum)
Srinagar	R	0.74 (4.58)	0.36 (2.38)	0.85 (5.65)	0.06 (0.40)	3.28 (21.64)	0.1395 (0.92)	1.6740 (11.05)
	U	0.42 (2.70)	0.07 (0.47)	3.45 (22.74)	0.08 (0.53)	4.14 (27.28)	0.0945 (0.62)	1.1340 (7.48)
	T	0.55 (3.49)	0.19 (1.27)	2.36 (15.60)	0.07 (0.47)	3.78 (24.92)	0.1133 (0.75)	1.3596 (8.97)
Anantnag	R	0.61 (3.94)	0.24 (1.55)	0.97 (6.25)	0.12 (0.72)	4.02 (25.96)	0.1212 (0.78)	1.4544 (9.41)
	U	0.99 (6.42)	0.16 (1.02)	0.89 (5.74)	0.19 (1.24)	4.08 (26.40)	0.1579 (1.02)	1.8948 (12.26)
	T	0.63 (4.09)	0.23 (1.52)	0.96 (6.22)	0.12 (0.75)	4.02 (25.99)	0.1235 (0.80)	1.4820 (9.58)
Baramulla	R	0.56 (3.56)	0.22 (1.38)	2.07 (13.14)	0.10 (0.63)	1.74 (11.08)	0.1195 (0.76)	1.4340 (9.12)
	U	0.54 (3.42)	0.06 (0.36)	1.72 (10.96)	0.09 (0.15)	3.30 (20.98)	0.0911 (0.58)	1.0932 (6.95)
	T	0.55 (3.54)	0.20 (1.25)	2.02 (12.85)	0.09 (0.63)	1.94 (12.39)	0.1158 (0.74)	1.3896 (8.84)
Kashmir	R	0.61 (3.91)	0.25 (1.64)	1.37 (8.80)	0.097 (0.063)	3.02 (19.46)	0.1240 (0.80)	1.4900 (9.60)
	U	0.48 (3.14)	0.07 (0.50)	2.94 (19.25)	0.091 (0.60)	3.99 (40.92)	0.1000 (0.65)	1.2000 (7.80)
	T	0.58 (3.72)	0.21 (1.35)	1.77 (11.44)	0.095 (0.62)	3.27 (24.88)	0.1200 (0.76)	1.4400 (9.14)

Note: Figures in parentheses are for per household
Source: Compiled from CORD data.

TABLE 4.11: **The Region-wise Distribution of Per Capita and Per Household Consumption Various Sources of Energy in Kashmir (in qts, per month, 1983-84)**

Dist.	Region	Fuelwood	Cowdung	Kerosene Oil	Charcoal	Cooking gas (kgs)	Total Energy (mtce)	
							Per annum	per capita p.m.
Srinagar	R	0.40	0.13	0.20	0.03	—	0.2529	0.0627
		(3.12)	(1.02)	(1.58)	(0.23)	—	(5.93)	(0.50)
	U	0.28	0.04	1.32	0.02	0.47	0.5942	0.0495
		(2.00)	(0.27)	(9.60)	(0.15)	(03.39)	(4.32)	(0.36)
	T	0.24	0.09	0.93	0.03	0.38	0.6607	0.0551
		(1.82)	(0.66)	(6.95)	(0.22)	(2.85)	(4.96)	(0.41)
Anantnag	R	0.34	0.16	0.11	0.02	—	0.7183	0.0599
		(2.63)	(1.28)	(0.86)	(0.17)	(–)	(5.65)	(0.47)
	U	0.36	0.06	0.42	0.008	0.01	0.2806	0.0234
		(2.57)	(0.43)	(3.03)	(0.06)	(0.10)	(2.04)	(0.17)
	T	0.34	0.15	0.16	0.02	0.025	0.6362	0.0530
		(2.65)	(1.13)	(1.26)	(0.16)	(0.19)	(4.93)	(0.41)
Baramulla	R	0.45	0.16	0.36	0.025	—	0.8834	0.0736
		(3.54)	(1.28)	(2.86)	(0.20)	(–)	(6.96)	(0.58)
	U	0.41	0.05	0.94	0.008	0.31	0.6630	0.0552
		(2.94)	(0.36)	(6.75)	(0.06)	(2.45)	(4.78)	(0.40)
	T	0.41	0.14	0.47	0.025	0.06	0.8379	0.0698
		(3.16)	(1.10)	(3.65)	(0.19)	(0.49)	(6.49)	(0.54)
Kashmir	R	0.37	0.16	0.25	0.025	—	0.7724	0.0644
		(2.89)	(1.25)	(1.97)	(0.19)	(–)	(6.08)	(0.51)
	U	0.24	0.03	1.29	0.025	0.64	0.5509	0.0459
		(1.76)	(0.24)	(9.39)	(0.18)	(4.66)	(4.00)	(0.34)
	T	0.33	0.12	0.58	0.025	0.21	0.7012	0.584
		(2.49)	(0.90)	(4.47)	(0.19)	(1.60)	(5.37)	(0.45)

Note: Figures in brackets are for per households.
Source: Compiled on the basis of Wood Balance Report.

TABLE 4.12: **Long-term Projection of Energy Consumption/Demand by Source in Kashmir (per month)**

Year	Region	Projected Population (lakhs)	Fuelwood ('000 qts)	Cowdung ('000 qts)	Kerosene oil ('000 lits)	Coal ('000 qts)	Electricity ('000 units)	Total Energy Demand	
								p.m (mtce)	per annum
1981	R	23.07	1407.15	576.70	3160.32	223.76	3896.67	286.43	3437.14
	U	8.28	397.49	57.97	2434.62	75.35	3304.12	82.73	993.72
	T	31.35	1804.64	634.67	5594.94	299.11	7200.79	369.24	4430.86
1986	R	25.76	1571.08	643.88	3528.50	249.83	7778.15	319.79	3837.56
	U	10.23	490.70	71.56	3005.53	93.03	4078.94	102.23	1226.75
	T	35.99	2061.78	715.44	6534.03	342.86	11857.09	422.30	5064.31
1991	R	28.45	1735.02	711.07	3896.67	275.90	8589.74	353.17	4237.98
	U	12.17	583.91	85.15	3576.45	110.70	4853.75	121.65	1459.77
	T	40.62	2318.93	796.22	7473.12	386.60	13443.49	474.81	5697.75

Source: Based on CORD data.

consumption per annum is 9.14 mtce (CORD) and 5.37 mtce according to wood balance study. There appears to be a considerable difference between these two studies even on the basic parameters.

Long- Term Forecasts

We have made predictions on energy consumption on the basis of two sources of data as referred to above and these forecasts are based on two methods as specified elsewhere. Table 4.12 reveals that the aggregate demand for fuelwood will rise from 180 lakh qts in 1981 to 206 lakh qts in 1986 and 232 lakh qts in 1991 (on the basis of CORD data). According to W.B. data the demand for fuelwood is likely to increase from 124 lakh qts in 1981 to 142 lakh qts in 1985 and 161 lakh qts in 1991. The demand for kerosene oil will go up from 559 lakh lits in 1981 to 653 lakh lits in 1986 and 747 lakh lits in 1991. The proportion of rural demand is much higher than the urban demand. The WBR puts the demand for kerosene oil at 282 lakh lits in 1991 and 250 lakh lits in 1985 (Table 4.13).

TABLE 4.13: **Long-term Requirements of Energy in Kashmir**

(W.B. Study based per year)

Year	Region	Fuelwood (lakh qts)	Kerosene Oil (lakh units)	Total Energy (lakh mtce)
1985	R	114.35	77.27	19.90
	U	29.44	158.25	5.63
	T	142.27	250.41	25.22
1991	R	126.29	85.33	21.98
	U	35.03	188.31	6.70
	T	160.80	282.63	28.46

The aggregate consumption/demand for energy per annum would go up from 443 lakh mtce in 1981 to 506 lakh mtce in 1986 and 570 lakh mtce in 1991 according to an analysis of CORD data. The demand for energy would be 25 lakh mtce in 1985 and 28.46 lakh mtce in 1991 when analysing W.B. data. (Tables 4.12 & 4.13). We have also made long-term forecasts by taking mean income elasticity of rural and urban sectors after analysing 28th Round of NSS Report.

Income Elasticity of Energy

The double log Engel functions were fitted to 24th and 28th Rounds of the NSS data and following fits were obtained:

R 24th Rd. $\log Y = -0.000003 + 0.3393\log X$

$R^2 = 0.99 \quad SE = 0.0036$

$Sig = .001$

U $\log Y = -0.000004 + 0.3788\log X$

$R^2 = 0.999 \quad SE = 0.0028$

$Sig = 0.001$

R 28th Rd. $\log Y = -0.000003 + 0.3972\log X$

$R^2 = 0.999 \quad SE = 0.0023$

$Sig = 0.001$

$\log Y = -0.000008 + 0.4344\log X$

$R^2 = 0.999 \quad SE = 0.0022$

$Sig = 0.001$

According to the Engel elasticity of 28th Round the demand for energy will increase to 308 lakh mtce in 1985 and 323 lakh mtce in 1991 (based on W.B. data). The demand for energy is likely to increase to 510 lakh mtce and 574 lakh mtce in 1985 and 1991 respectively if we rely on CORD data.

If we take double exponent the demand for energy is likely to increase by 530 lakh mtce in 1985 and 624 mtce by 1991 and if we rely on CORD data and 320 lakh mtce in 1985 and 377 lakh mtce in 1991 per annum when W.B. data is analysed (Table 4.14).

Conclusion

We would not like to present a formal conclusion with policy options here as the policy options cannot be identified unless the data base is reliable.

However, we feel that the cost of electricity per unit on an average works out to 27 paise and the hydro-power generation appears to be the cheap alternative in the State. It is felt that the losses are considerable which can be minimized by technical and management efficiency. In the light of growing income, population and urbanization the existing potential needs to be effectively utilised. Regarding traditional household consumption of energy, the propensity to consume fuelwood continues to be high which can act as a drain on rich forest resources. The cheap alternative sources need to be explored in view of the rising demand. The existing energy efficiency also needs to be raised to optional level.

TABLE 4.14: **Long-term Projections of Energy Demand in Kashmir**

('000 mtce)

		CORD data		W.B. data	
Year	Region	per month	per year	per month	per year
1981	R	280.58	3462.98	179.70	2156.40
	U	83.63	1003.65	46.04	552.48
	T	364.21	4466.63	225.74	2708.88
1985	R	322.20	3866.40	200.63	2407.56
	U	103.25	1239.00	56.84	682.08
	T	425.45	5105.40	257.47	3089.64
1990	R	358.82	4269.84	201.66	2419.92
	U	122.86	1474.37	67.64	811.68
	T	478.68	5744.21	269.30	3231.60
1985*	R	334.05	4008.58	207.85	2496.16
	U	107.34	1288.08	59.19	710.28
	T	441.39	5296.66	262.04	3204.44
1990*	R	385.69	4628.22	240.06	2880.70
	U	134.05	1608.67	73.84	886.08
	T	519.74	6236.89	313.90	3766.79

Notes: mtce stands for metric tonnes of coal equivalent. The years without star mark are based on income elasticity exponent only.
The predictions for the years with * mark are compounded with exponents time and income elasticity.
CORD refers to Centre of Research for development, University of Kashmir.
W.B. refers to Wood Balance data.

NOTES

1. Parikh, R. (1980): *Energy Systems and Development,* Oxford Univ. Press, New Delhi.
2. World Bank (1976): *Report on Energy Demand Forecasting: A Critical Review of Current Approaches,* No. 229, Washington.
3. Henderson, P.D. (1983): "India's Energy Problems" in *Energy Problems and World Economic Development* with Special Reference to India.

4. Singh, R.R. and Gaur, A.K. (1983): India's Energy Crisis: A Survey of Viable Alternatives," (Ed.) in *India's Energy Problem.*
5. N.C.A.E.R., Survey on Domestic Energy Consumption.
6. *Ibid.*
7. Singh, R.R. and Gaur, A.K. *op. cit.*, 1983
8. *Op. cit.*
9. Singh, C.D. and Singh, Y.P., (1983): "India's Energy Crisis: Some Policy Options (ed.) in *India's Energy Problem.*
10. Directorate of Economic & Statistics, J&K Govt, Srinagar, 1977-78: *Digest of Statistics.*
11. *Ibid.*, 1984-85: *Draft Economic Review.*
12. *Ibid.*, 1983-84: *Draft Economic Review.*
13. *Ibid.*, 1984-85: *Economic Review.*

FIVE

FUELWOOD CONSUMPTION PATTERN AND ENVIRONMENTAL DEGRADATION IN KASHMIR
A Case Study of Naranag Catchment

MAJID HUSAIN

The relentless increase in the demand of fuel and power raises important questions about the future security of the world energy supply. Energy consumption and the consequent stress on fuel resources is accelerating from year to year, and there are inter-regional and intra-regional variations in the fuel and energy demands. The problem has been accentuated by the tremendous increase in population, the industrialization, urbanization and growing investments in the living standards of people. The crisis in energy – its misuse, overuse, and underuse, is, however, more acute in the developing countries. India in general and several of its meso-regions like Jammu and Kashmir in particular are confronted with the scarcity of developed energy resources. The inadequacy of power has been affecting adversely the pace of development in all sectors of economy of Jammu and Kashmir State.

In India the conventional fossil fuel i.e. coal, kerosene oil, natural gas is generally consumed by the urban dwellers, while in the rural areas, people almost exclusively depend on cowdung and fuelwood for cooking, heating and domestic lighting. The dependence of *Kandi* people on forests for fuelwood is strikingly significant. In the undulating and relatively isolated rimlands of Kashmir fossil fuel is not easily available. Moreover, the poor masses cannot afford to purchase expensive coal, kerosene and gas. In the high altitudinal low-pressure zone of inclement weather, the mild heat of cowdung is ineffective fuel for cooking and heating. In some of the clans of Kashmiri Gujjars who dwell in *Kandi* areas, use of cowdung for cooking is a social taboo. Having their habitat in the forested slopes, the rural masses of hilly tracts are naturally tempted to use fuelwood for cooking, heating and other diversified social needs. Under such a geo-ecological setting the forests are being misused and indiscriminately exploited.

The present paper is an attempt to examine the existing patterns of fuelwood consumption in the *Kandi* areas of Kashmir and to ascertain the economic loss and ecological implications of excessive fuelwood consumption in the highly vulnerable side valleys of Kashmir Division. For this purpose, the Naranag Catchment – a right-hand tributary of the Sindh river – has been selected as the area of study.

The study is based on the data generated by the author in March, June and October, 1984. Apart from field observations, data and information were collected in the form of structured village and household schedules.

Naranag meets the Sindh river at the village of Kijipura about two kms down to the town of Kangan. This subsequent tributary of the Sindh river presents a panoramic view, characterized by perennial streams, brooks, fresh air and diverse fauna and flora. In the upper reaches of Naranag are two settlements of the Gujjars, namely Pahalnar and Naranag which have been taken as the samples for study.

Climatically, Naranag Catchment has a marked seasonality with severe snowy winters. After September the night temperature generally slumps below the freezing point, while the average January temperature ranges between 2.8°C and 3°C. About six months (October to March) are the months of extreme cold, and it is during these months that the consumption of fuelwood for cooking and heating is maximum. April-to-May is the period of warm sun with intermittent rains from the "Western Disturbances". The inclement weather of winters forces the people to remain inside their shelters (*Kothas* and *Bandis*) and they sit near *Dens* (indigenous stones) to protect themselves against cold. Use of *Kangri* (fire-pot) is uncommon amongst the Gujjars of Naranag and Pahalnar villages. Consequently, the use of the fuelwood for cooking and heating in indigenous stoves *(Dens)* is enormous.

Naranag and Aharbal villages – located on the right bank of the Nara stream have 49 and 51 households respectively. Each household of these villages was surveyed for gathering information about fuelwood consumption. The total population of Naranag and Aharbal as per the enumeration in January 1985 was 296 and 344 respectively. Though the sites of these villages have archaeological imprints, the present settlements are relatively new and about 60 per cent of their households came into existence only during the last three decades. In fact, the well-off Gujjars, who practised transhumance between the slopes of the Shiwaliks (Jammu Division) and the *Margs* (Alpine pastures) of the Gangabal tract, started developing land connections in the early sixties on the slopes of Naranag. The traditional shepherds (Gujjars) after developing land connections, have changed their mode of life and adopted the occupation of farming, thereby getting closely attached to soils and forests instead of pastures.

The inhabitants of Naranag and Aharbal villages have joint family system. The average size of a family is 6.75 in Pahalnar and 6.10 in Naranag. Interestingly enough, out of the 100 families of these villages only 26 households have less than four persons, 25 households consist of five to six members and 30 families are based on seven to eight members, while 19 households have 10 to 14 members (Table 5.1). The sex ratio is 858 per thousand which is fairly below the State and National levels. Forty-four per cent of the total population constitutes the work force which is engaged mainly in primary activities.

TABLE 5.1: **Size of Family — Naranag and Pahalnar Villages (January 1985)**

(*Number of Households*)

Size of Family	Number & Percentage of Households
Below 4 persons	26
5 - 6 persons	25
7 - 8 persons	30
9 -10 persons	10
More than 10	9
Total	100

Fuelwood Consumption Pattern

In the Naranag Catchment, it has been observed that each household, irrespective of its family size, workforce and dependency ratio, almost entirely depends on fuelwood for cooking and heating purposes. The level of consumption of fuelwood is, however, significantly influenced by household size, age structure and the nearness of forests. Petroleum gas, coal, lignite, charcoal, kerosene and cowdung as discussed at the outset are not used for cooking and heating. Consequently, fuelwood is the dominant source of energy in the Naranag catchment. Under the existing geo-ecological and socio-economic conditions the people of the area have very little alternative to fuelwood.

An average picture of the annual fuelwood consumption in the families of selected villages has been given in Table 5.1. It may be seen from this Table that on the average a family consumes nearly 54 quintals of fuelwood annually. A small household, having less than four family members, needs about 47 quintals of fuelwood, while the large sized families having 10 or more persons need about 58 quintals of fuelwood for cooking and heating.

An interesting feature of the consumption pattern of fuelwood is that the per head consumption of fuelwood in the small sized families is higher (about 12 quintals per annum) to that of large sized families in which the per head consumption of fuelwood is only 4.48 quintals per annum, i.e. almost one-third of that of small families. It may also be noted from Table 5.2 that there is a gradual decrease in wood consumption as the size of family increases. Although it is difficult to ascertain the percentage share of fuelwood consumed in cooking and heating, nevertheless, the processed data reveals that about 45 per cent of the total fuelwood is required for cooking and 51 per cent for heating, while about 14 per cent for lighting purposes.

In order to ascertain the household fuelwood consumption pattern, the physical quantities of wood have been converted into British Thermal Units (BTU) equivalents. The conversion factors used in the conversion of physical quantities of different fuels into British Thermal Units are given in Table 5.3.

The energy consumption level for cooking and heating obtained from fuelwood and other sources in the villages under study has been given in Table 5.4. The fuelwood and charcoal used by the villagers have been converted into the British Thermal Units. The main advantage of the conversion lies in the fact that the results thus obtained may be compared with the national and international levels.

An examination of Table 5.4 shows that the mean monthly consumption of fuelwood and charcoal for cooking, heating and lighting per household in the Naranag and Pahalnar villages is about 90.58 million British Thermal Units. In per capita terms this represents about 12.63 million BTU per month. The mean monthly fuelwood consumption per household for cooking purposes in the catchment was found to be about 41.48 million BTU per month as against the national

TABLE 5.2: **Per Annum Fuelwood Consumption in Naranag Catchment**

(Figures in Quintals)

Size of Family	Cooking		Heating		Lighting		Total	
	Fuel-wood	%	Fuel-wood	%	Fuel-wood	%	Fuel-wood	%
Less than 4 persons	20.00	44.35	25.00	53.19	2.00	4.26	47.00	100.00
5-6 persons	23.00	44.35	26.66	51.40	2.20	4.24	51.86	100.00
7-8 persons	25.00	45.58	27.50	50.13	2.35	4.28	54.85	100.00
9-10 persons	26.00	45.83	28.33	49.94	2.40	4.23	56.73	100.00
More than 10 persons	26.50	45.56	29.16	50.13	2.50	4.30	58.16	100.00
Average	24.10	44.86	27.33	50.87	2.29	4.26	53.72	100.00

TABLE 5.3: **Factors for Converting Quantities to Energy Equivalents**

1 kg. firewood	=	14259 BTU
1 kg. charcoal	=	28099 BTU
1 ltr. kerosene	=	36745 BTU
1 kg. LPG (liquid petroleum gas)	=	43569 BTU

average of 2.1 million BTU and that of Hyderabad City 2.5 million BTU. Thus the per-head consumption of fuelwood for cooking, heating and lighting in almost 75 times of that of the national average. Under the low-pressure conditions of the area, more heat is required for cooking of food, even then enough fuelwood can be saved if better utensils and *Dens* (indigenous stoves) are used.

TABLE 5.4: **Fuelwood Consumption Pattern in Naranag Catchment**

(*Figures in BTU*)

Size of family	Fuelwood for cooking	Fuelwood for heating	Use of wood for lighting	Use of charcoal per month	Total	Average per head consumption
less than 4 persons	30.79	42.77	2.85	0.07	76.48	19.12
5-6 persons	39.21	45.62	3.14	0.08	88.05	14.68
7-8 persons	42.77	47.05	3.35	0.09	93.26	11.65
9-10 persons	44.20	48.48	3.42	0.11	96.21	9.62
More than 10 persons	45.62	49.90	3.26	0.12	98.90	8.08
Average	41.48	46.76	3.20	0.09	90.58	12.63

Source: Field work by the author.

Apart from cooking, fuelwood is also required in substantial quantities for heating especially during the winter months. In winters the outdoor activities remain suspended and the family members sit near the *Dens* to combat cold. On an average, each family, irrespective of its size, consumes about 25 to 29 quintals of fuelwood for heating. In winters the supply of electricity is irregular and load shedding is a regular feature. Under such conditions rural population uses oil-bearing wood of *Kail* for light in *Kothas* and *Bandis*.

It may also be noted from Table 5.4 that in each household about 98 per cent of the total energy is obtained from fuelwood and the remaining 2 per cent is acquired from charcoal. Use of coal, petroleum gas and electricity for cooking and heating was not reported from any household of the selected villages.

The fuelwood used in the *Kandi* and rural areas of Kashmir is obtained mainly from *Budloo, Deodar* and *Kail* forests. The wood of these evergreen species has great commercial value if sold as timber and timber products. Information on the consumption of various types of wood, twigs, branches and scrubs and the rate of fuelwood and timber in the neighbouring markets was obtained on the basis of which Table 5.5 has been computed.

TABLE 5.5: **Value of Fuelwood used in Aharbal and Naranag Villages**

(Total population 640)

Fuelwood used	Fuelwood consumed (in qts)	Price per quintal (in Rs.)	Amount (in Rs.)	Price of commercial timber (in Rs.)	Amount (in Rs.)	Loss (in Rs.)
Budloo	667.2	20	1,3344	115	76,728	63384
Kail	1,175.2	25	2,9380	150	176,280	146900
Deodar	840	30	2,5200	180	151,200	126000
Twigs, branches & other wood	3,091.6	15	4,6374	15	46,374	—
Total	5,774	-	11,4294	-	450,582	336284

Source: Field work by the author.

An analysis of Table 5 shows that the total consumption of fuelwood in the sample villages having a population of 640 is about 5,774 quintals per annum. The cost of the fuelwood consumed has been assessed at the rate of Rs. 20 per quintal for *Budloo*, Rs. 30 for *Deodar*, and Rs. 25 for *Kail*. The total cost of all varieties of wood and twigs and branches consumed annually is thus estimated to be Rs.114,294. The superior quality of wood used as fuel is about 58 per cent of the total wood consumption, while the remaining 42 per cent is obtained from twigs and scrubs. If the conifer wood is converted into timber and sold in the neighbouring markets of Kangan and Srinagar, the wood consumed annually may fetch about Rs.450,582. The gross loss resulting from the careless burning of valuable wood is 336,284. If such an exercise is carried on for the entire State, the loss is colossal and this needs immediate attention of researchers and policy makers.

Fuelwood Consumption and Environmental Degradation

As stated at the outset the number of households in Naranag and Pahalnar villages has increased from 20 and 21 to 49 and 51 respectively over the past 30 years. This unprecedented increase is the result of land connections and sedentry style of life of the Gujjars who were previously practising transhumance. The growing demand for cereals, edibles, goods, timber and fuelwood has put an additional pressure on the limited arable land of the selected villages. Consequently, there is a significant encroachment into forests. Steep slopes have been brought under plough. The associated demand on resources, technology and the interaction among these factors is becoming increasingly consequential. Some of the adverse effects observed in the area of study as a result of heavy consumption of fuelwood by the inhabitants and over utilization of forests are given in brief in the following lines.

The new houses are being constructed on steep slopes ranging between 30° and 45°. Once a house is constructed the adjacent forest is lopped off and trees felled indiscriminately to develop land for maize cultivation. This development has accelerated the process of soil erosion which is the most dramatic aspect of human induced modification of the beautiful side valley of Naranag. Although detailed investigations about the extent of erosion are yet to be made, it is so widespread and of such serious proportions that the visitors and field workers cannot fail to notice it. The development of agricultural land on steep slopes and growing soil exhausting crops like maize in the thin coarse alluvial soil have made the eco-system of the region more fragile which has lost its resilience characteristics. In fact, the new farmers (Gujjars) are not very efficient in soil management.

The excessive consumption of good quality wood for fuelwood and conversion of timber into charcoal is not only making the eco-system less resilient, there are serious indirect adverse effects. The deforestation is bringing chemical and physical changes in the natural surroundings of the area. The newly sedentrized farmer, while developing a new patch of land for agriculture, set the forest on fire. This practice has far-reaching consequences. For example, when a new settler fells trees and ignites them to clear land for settlement and arable lands, the temperature of the top soil for a few minutes may soar to over 60°C. Consequently, the soil micro-flora and micro-fauna including decayed organic matters are destroyed. More subtle but probably of greater significance is the change in soil temperature that results when forest cover is removed and the soil is exposed to sun and other agents of weathering and denudation. In a forest soil temperatures tend to be nearly isothermal, i.e. they experience only slight diurnal range of temperature. The high diurnal range of temperature accelerates the process of mechanical weathering resulting in depletion of soils. In several localities of the Naranag Catchment, especially in the vicinity of Aharbal and Naranag villages, one may observe the leached and barren rocks giving the look of ecological slums. Climatic data for such a micro-region is not available, nevertheless, the elderly people of the area complain that precipitation is on the declining side and the general weather conditions are changing.

Forests, which have been indiscriminately felled for fuel and timber, play a useful role in the maintenance of ecological diversity, preservation of water heads, and prevention of erosion, moderation of climate, supplying wood for fuel, structures and papers, and providing hunting grounds and areas of aesthetic value for recreational purposes. In brief, deforestation leads to soil erosion and makes run-off more erratic, reducing availability of water and causing water pollution. This process can become irreversible by altering the environment so drastically that the eco-system may lose its resilience characteristics, making afforestation impossible.

In the Naranag catchment, if the process of sedentrization, and excessive fuelwood consumption are not checked, we can expect frequent and more serious environmental catastrophe in the form of changes in climate, sterility of soil, and water and air pollution. The problem has become acute in the valley, the solution of which cannot be delayed further.

REFERENCES

Alam, S.M., Dunkerly, J. *et al*, (1984), *Fuelwood in Urban Markets — A Case Study of Hyderabad*, Concept Pub. Co., New Delhi.

Bonnett, C.F., (1975), *Man and Earth Eco-systems*, London.

Borgstorm, G., (1969), *Too Many*, Macmillan, New York, p. 72.

Ehrlich, P.R., et al., (1977), *Ecoscience*, W.H. Freeman & Co., Sanfrancisco.

Husain, M., (1984), "Landscape Ecology — Some Concepts," in *M.R. Chaudry, Feliciation Volume*, pp. 49-58.

Routley, R. and V. Routley, (1975), *The Fight For the Forests*, Australian National University, Canberra.

Spurr and Barnes, B.V., (1973), *Forest Ecology*, Ronald, New York.

SIX

FUEL CONSUMPTION PATTERN IN SRINAGAR MUNICIPALITY
A Spatial Analysis

BHUPINDER ZUTSHI & MONDIRA DUTTA

Energy is an essential input for all economic activities. The economic condition of a country or a region can best be judged on the basis of its level of energy consumption. It is a pity that even in the present times when energy resources like hydro carbon, bio-gas and solar energy are being successfully utilized, firewood continues to be the major fuel resource in the valley of Kashmir. And the dominance of fuelwood is not confined to the rural hinterland only but paves its path among the elite urban centres of the region.

It is needless to emphasize that there has been a tremendous increase in the demand for fuel resources during the past one decade, especially in the urban nodes of the country. Thus in the absence of a proportionate increase in the supply has unfortunately resulted in the black-marketing of particular fuel resources, one of them being firewood. One of the major reasons for the depletion of forest resources, leading to ecological imbalances, could be attributed to the large-scale consumption of firewood. A study of the fuel consumption pattern is, therefore, of crucial importance, since it would not only lead to an equitable distribution of fuel resources but also highlight the consumption pattern and the future demand.

Such a study of the Valley of Kashmir can be analysed with the help of the traditional sources of energy, i.e. firewood, kerosene oil, cowdung, coal, gas and electricity. The present paper is an attempt to examine the consumption pattern of the various alternative sources of energy in Srinagar Municipality which constitutes the major urban centre of the Kashmir division. The study, however, has not taken into account the fuel resources like coal, hydro-electricity and cowdung. Coal is very expensive as it is not mined within the valley, information about supply of electricity is not reliable, and the use of cowdung is confined to rural areas. Owing to these reasons their consumption pattern has been excluded from the present study.

In order to ease the pressure on firewood and kerosene, supply of petroleum-gas was started in the valley in 1967. The consumption pattern of kerosene oil and cooking gas in the valley has been given in Table 6.1.

TABLE 6.1: **Consumption Pattern of fuel (Kerosene Oil and Cooking gas) in Kashmir Valley**

Sl. No.	Spatial Unit	% of Population	% of Kerosene oil	% of Gas connections
1.	Srinagar Municipality	18.70	66.76	82.11
2.	Anantnag Distt.	20.92	9.13	4.87
3.	Baramulla Distt.	21.38	9.64	13.92
4.	Kupwara Distt.	10.46	2.84	—
5.	Pulwama Distt.	12.91	7.31	—
6.	Badgam Distt.	11.26	2.82	—
7.	Srinagar (Rural)	4.33	1.50	—
	Kashmir Valley	3130090 (100%)	2592150 (100%)	6150 consumers

Sources: 1. Food and Supplies Department, Srinagar, and
2. Jaykay Gas Company, Srinagar.

An examination of Table 6.1 shows that the Municipal region of Srinagar is privileged in the matter of kerosene supply and gas connections. It is interesting to note that the Srinagar Municipal area having 19 per cent of the total population has 67 per cent of the total kerosene oil and 82 per cent of the total gas supply of the valley. Another interesting point to be noted is that despite the large-scale consumption of kerosene oil and gas has not led to a declining trend in the consumption of firewood in Srinagar. If such be the situation in the Srinagar Municipal area where a variety of fuel energy resources are available one can visualize the plight of the rural areas where the major toll is on firewood alone. Therefore, it is quite possible that such a reason is playing a major role in the deforestation process. It is important to note from Table 6.1 that the Srinagar Municipal area constituting 18.70 per cent (585,464 persons) of the total population consumes 82.11 per cent (5,050 consumers) of the total gas. This clearly reflects that hardly .008 per cent of the total population of Srinagar Municipality has been extended the privilege of gas supply, the rest being dependent mainly on firewood and kerosene oil. Therefore, there still seems a scarcity of fuel resources in Srinagar Municipality. The large-scale immigration from the rural areas may partly be held responsible for such a dearth.

Objectives of the Study

The study has the following objectives:

(i) To analyse the spatial distribution of firewood, kerosene oil and gas supply depots in the Municipality of Srinagar;
(ii) To study the consumption pattern of the selected fuel resources in the wards of Srinagar Municipality;
(iii) To analyse the temporal variations in the consumption pattern of the fuel resources under study; and
(iv) To ascertain the imbalances in the consumption pattern of fuel resources.

Hypotheses

The present paper has the following three hypotheses:

(i) The consumption of firewood is the highest amongst the selected fuel energy resources during winter i.e. December to March;
(ii) The per head consumption of firewood is less in the areas where consumers of high income group reside;
(iii) The wards having low annual consumption of kerosene will show a high annual consumption of firewood per head.

Data Base

The entire data base for the study has been collected from the following sources:

(i) The firewood section of the Forest Department, Srinagar
(ii) Food and Supplies Department, Srinagar;
(iii) Directorate of Census Operations, Srinagar; and
(iv) Town and Country Planning Organization, Srinagar.

Methodology

The study utilized the statistical techniques of Gini's Coefficient and Lorenz Curve to find out the uneven distributional pattern of fuel resource in the municipal area of Srinagar. Other basic tools like percentage and growth rates have also been used. The results arrived at have been plotted in maps by applying the cartographic techniques of choropleth, piediagrams, bar diagrams and graphs.

The Study Area

Srinagar city forming the core of the Kashmir region is centrally located on both the sides of Jhelum river in the vicinity of the Dal Lake. In terms of population and area it is the largest city in the State of Jammu and Kashmir. Srinagar has functioned as the regional focal centre due to its advantageous and strategic location. The city has grown rapidly over the past two decades due to the concentration of economic activities, administrative activities and educational institutions. The growth rate of the city over the period of past few decades has been given in Table 6.2.

TABLE 6.2: **Decadal growth rate of population of Srinagar Municipality**

Year	Total Population	Decadal Variation
1911	126344	+3.04
1921	141734	+12.18
1931	173573	+22.46
1941	207787	+19.71
1951	246522	+18.64
1961	285257	+15.71
1971	403413	+41.42
1981	585464	+42.89

Source: Census of India, Jammu and Kashmir Population Total 1981, Series 8.

Table 6.2 shows a very high decadal growth rate of population during 1961-71 and 1971-81. This stupendous growth rate has resulted in the large-scale demand of fuel resources in the area.

The city of Srinagar consists of 17 wards constituting a total area of 82.88 sq. kms. Out of the 17 wards, eight (Nos. 6, 7, 8, 9, 10, 11, 13 and 14) are located in the core areas along the sides of river Jhelum. The population of the area is mostly engaged in the service sector or the small-scale trade and commerce sector as it is the hub of economic activities. Six wards numbered 1, 3, 4, 15, 16 and 17 are located in the peripheral areas of Srinagar municipality. These wards are the recent outgrowth of the Srinagar municipal area and are mostly devoted for residential use and large-scale industrial activities. The remaining three wards numbered 2, 4 and 5 have a large concentration of industry especially large-scale, like the Silk Factory, Joinery Mills, Electronics and the H.M.T. watch factory.

Distribution of Registered Consumers

The percentage of registered consumers to total population for the energy resources like firewood and kerosene oil records 67 and 76 respectively as shown in Table 6.3. The percentage proportion of registered consumers in the case of gas supply is rather dismaying recording hardly .008 although Srinagar Municipal area alone consumes as much as 82 per cent of the total gas supply in the valley.

It may be seen from Table 6.3 that gas consumers mostly belong to ward numbers 1, 2, 3, and 9. Most of the wards are located in areas where the residents are either rich businessmen or are high-ranking civil servants. The old city area having large population, has a negligible percentage of gas consumers.

There is a marked variation in the registered firewood consumers in the different wards. More of the registered consumers are concentrated in the core areas of Srinagar Municipality. These wards are numbers 2, 3, 4, 6, 7, 8 and 10. From Table 6.3 it is clear that ward numbers 1, 13, 15, 16 and 17 have a comparatively low percentage of fuelwood consumers. Most of these are the newly settled localities which either do not use fuelwood at all or buy it directly from the contractors.

The percentage of consumers registered for kerosene oil on the other hand shows a fairly equitable distribution among the wards. Ward numbers 2, 3, 7, 10 and 14 record cent per cent registered consumers for kerosene oil. Even the lowest percentage of registered consumers in any ward is as high as 75 (ward number 11). This clearly indicates that kerosene oil is much more popular as compared to firewood in the city.

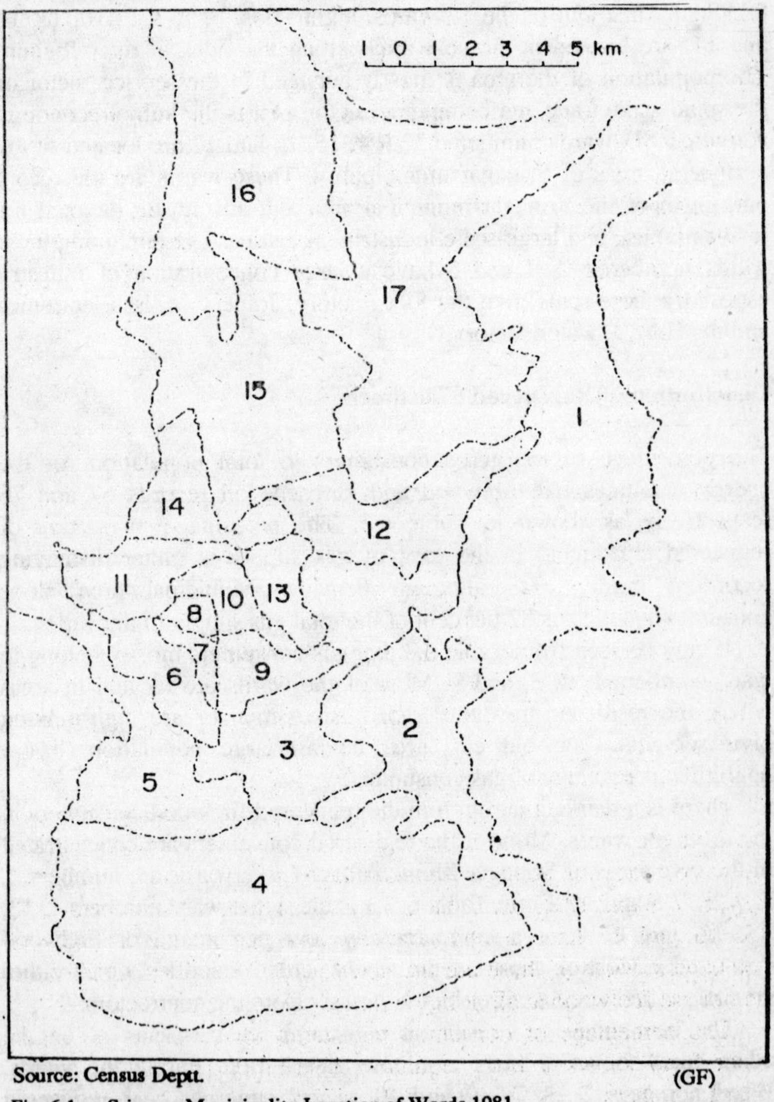

Source: Census Deptt. (GF)

Fig. 6.1: Srinagar Municipality Location of Words 1981

TABLE 6.3: **Percentage of Registered Consumers to Total Population**

Sl. No.	Ward No.	Name of Area	Total population	Fire-wood	Kerose-ne oil	Gas
1	I	Nishat	29951	40	78	1
2.	II	Batwara	19179	94	100	1
3.	III	Amirakadal	23237	94	100	3
4.	IV	Barzalla	45333	92	78	.80
5.	V	Batmaloo	46916	77	76	.08
6.	VI	Chattabal	42830	90	93	.08
7.	VII	Kanikadal	24244	92	100	.01
8.	VIII	Nai Sarak	29738	91	88	.01
9.	IX	Habbakadal	43957	59	76	3
10.	X	Maharaj Gunj	44669	89	100	.80
11.	XI	Safakadal	37558	49	75	.04
12.	XII	Rainawari	43221	52	83	.003
13	XIII	Zainakadal	36837	38	83	.004
14.	XIV	Zadibal	30672	68	100	.004
15.	XV	Lal Bazar	32644	41	83	.004
16.	XVI	Soura	25987	39	81	.003
17.	XVII	Hazratbal	28491	41	91	.003
	Total		585464	67	76	.008

Sources: 1. Directorate, Census Operations, Srinagar.
2. Jaykay Gas Company, Srinagar.
3. Directorate of Food and Supplies, Srinagar.

Distribution Pattern of Energy Depots

A Table 6.4 has been prepared to present the distribution pattern of the various energy resource depots. Table 6.4 shows the percentage of depots for particular energy resource to the total mohallas as well as to the total population of a ward.

An examination of Table 6.4 shows that only 11 per cent and 26 per cent of the total *mohallas* have been provided with firewood and kerosene oil depots respectively. The percentage of depots for firewood is the highest in ward numbers 4, 5 and 6 and the lowest is in ward numbers 1, 13, 16 and 17. The population per depot has a close association with this. These wards where the percentage of depots to mohallas is high records a low population per depot. This suggests that the higher the percentage of depots the lower will be the population threshold. The pressure of population per depot naturally becomes less. On the other hand those wards where the percentage of existing depots is low, the population threshold becomes high, showing that a large number of people are dependent on the depot.

The picture for kerosene oil is a little balanced as compared to firewood in the sense that most of the wards show a uniform pressure of population per depot.

An interesting feature to be noted is that the wards where the population pressure is high for kerosene oil (namely, ward numbers 1 and 13) also records a high population pressure for firewood as well. This clearly shows that the overflow of population shifts from kerosene oil to firewood probably because of a shortage in kerosene oil. This excessive pressure could be one of the factors responsible for the people not receiving their kerosene oil in time.

The distribution of consumer gas depots is highly uneven, being located at a few places in the Municipality. Six depots namely Lal Chowk, Hari Singh High Street, Karan Nagar, Nai-Sarak, Khanyar and Zainakadal are located in the vicinity of the main business centre as well as in the core limits of the city while the remaining three (Durganag, Lalmandi and Barzalla) are serving the other hinterlands of the municipality. The distribution, however, reflects a highly random distribution whereby many chunks of the area are not served.

Distribution of Fuel Resources

Table 6.5 shows the temporal variations in the distribution of energy resources for 1982 to 1984, seasonwise. At a glance Table 6.5 gives a rather optimistic picture showing a decline in the consumption of firewood and an increase in the growth rate of kerosene oil and gas consumption. But the magnitude of decrease in the distribution of firewood is marginal as compared to the increase in kerosene oil and consumer gas distribution pattern.

The entire year has been divided into three seasons consisting of a period of four months each, beginning from December onwards. Almost in all the fuel resources except for kerosene oil the distribution is quite high during the winter period, i.e. December to March. Especially in the case of firewood the distribution is very high during the winter season. This is so, because there is a dearth of kerosene oil in the valley during this time due to the constant road blockade. Moreover, firewood is mostly used for internal heating purposes. Therefore, the consumption is high during winters. The firewood distribution is the minimum during summers, i.e. April to July. From August to November, there is a steady rise in the demand of kerosene, most probably owing to the storing tendency of people for the winters.

TABLE 6.4: **Distribution of Energy Depots**

Sl. No.	Ward No.	Total Mohallas	Firewood			Kerosene Oil			
			No. of Depots	% of Col. 4 to Col. 3	Population per depot	No. of Depots	% of Col. 7 to Col. 3	Population per Depot	No. of Depots
1.	I	23	2	9	14576	2	9.00	14576	—
2.	II	19	3	16	9393	10	53.00	1917	1
3.	III	31	4	13	5809	10	32.00	2323	3
4.	IV	27	8	30	5666	19	70.00	2385	1
5.	V	40	9	23	5213	12	30.00	3893	—
6.	VI	25	5	20	7420	11	44.00	3373	1
7.	VII	22	4	18	6061	12	55.00	2020	—
8.	VIII	28	4	14	7435	12	43.00	2478	1
9.	IX	38	4	11	10989	20	53.00	2197	—
10.	X	52	7	13	6381	8	15.00	5584	1
11.	XI	37	4	11	9389	10	27.00	3755	—
12.	XII	89	4	5	10805	6	27.00	7203	—
13.	XIII	46	2	4	18418	4	9.00	9209	1
14.	XIV	42	4	10	7668	10	24.00	3067	—
15.	XV	43	3	7	10881	9	21.00	3672	—
16.	XVI	24	2	8	12993	3	13.00	8662	—
17.	XVII	37	2	5	14246	4	11.00	7123	—
	Total	623	71	11	8246	162	26.00	3613	9

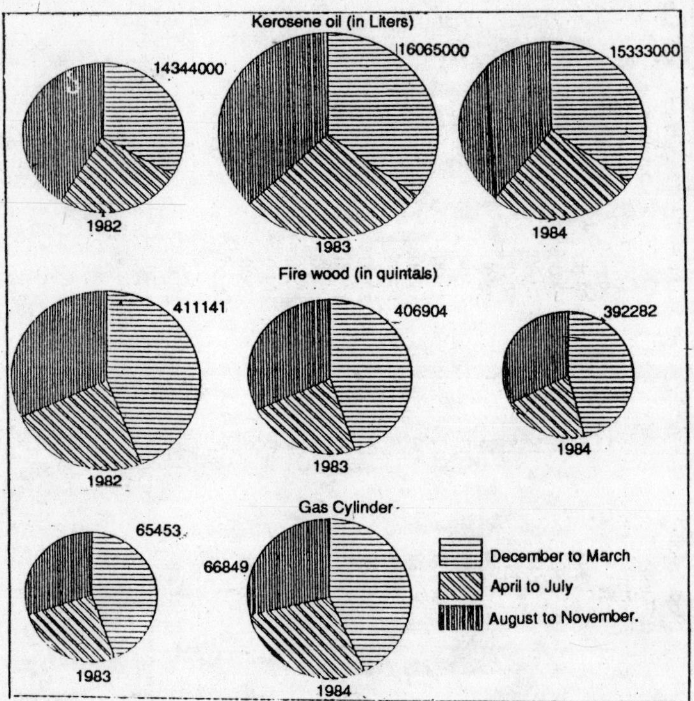

Source: (1) Food & Supplies Department Srinagar. (GF)
(2) Forest Department Srinagar.
(3) Vay Kay Gas Company Srinagar

Fig.6.2: Srinagar Municipality Fuel Energy Distribution 1982-84.

The kerosene oil percentage distribution shows a more or less stable pattern throughout the year. The variations are not as glaring as in the case of firewood. The distribution is the highest during autumn which is normal keeping in view that winter is not far away. And there is a scarcity of kerosene oil during winters, so the maximum storage is done during autumn. Perhaps the scarcity during winters leads to a comparatively low percentage distribution of kerosene oil between December and March. The percentage distribution during summer is also lower than during autumn as the urgency of storing the fuel is not so immediate. The percentage distribution of consumer gas like firewood is high during winter. The summer and autumn have a more or less similar distribution with a scarcity of kerosene oil while firewood being the only dependent fuel during winters, the gas distribution does gather momentum in this period.

Consumption Pattern of Fuel Resources

A wardwise distribution of High Income Group (HIG) registered consumers depicts large-scale variations among the wards of Srinagar city. Ward No. 1 records a very high percentage (more than 50) of HIG registered consumers. This ward is the newly developed area in Nishat. This is closely followed by ward numbers 3, 4, 5, 9, 16 and 17 (30 to 38 per cent). All those wards correspond to the outer zone of Srinagar city, which are quite recently developed residential areas. Most of these wards are thronged with rich business class people or high-ranking civil servants who have moved out from the congested core areas. A low percentage of HIG registered consumers belonging to ward numbers 6, 8, 11, 13, 14 and 15 are mostly from the old city areas.

The fuel consumption in terms of firewood is very high in ward numbers 3, 6, 7, 8, 10, 12 and 13. These wards are all more or less located in the core areas of the city where a large chunk of the Low Income Group (LIG) families reside. This clearly establishes a close and positive association between the LIG consumers and firewood consumption. The reverse is also true in the sense that, all the wards of the newly developed area, where a large section of HIG consumers reside, show a low level of firewood consumption (ward numbers 1, 15, 16 and 17).

The consumption pattern of kerosene oil is more uniform among the wards. However, ward no. 3 (Lal Chowk area) which is the main business centre reflects a very high annual consumption of kerosene oil per head (41 lits) as compared to the norm of 26 for the city. This could be attributed to the norm or presence of large-scale hotels. The table also reflects the wards having a very low annual consumption of kerosene (Ward Nos., 7, 8, 10 and 13) show a very high annual consumption of firewood. In the case of Ward No. 1 (Nishat area) such association was not observed as both per-head annual consumption of kerosene oil and firewood was very low. This is probably because the consumers being rich have a high purchasing power and hence may be purchasing firewood directly from the contractors.

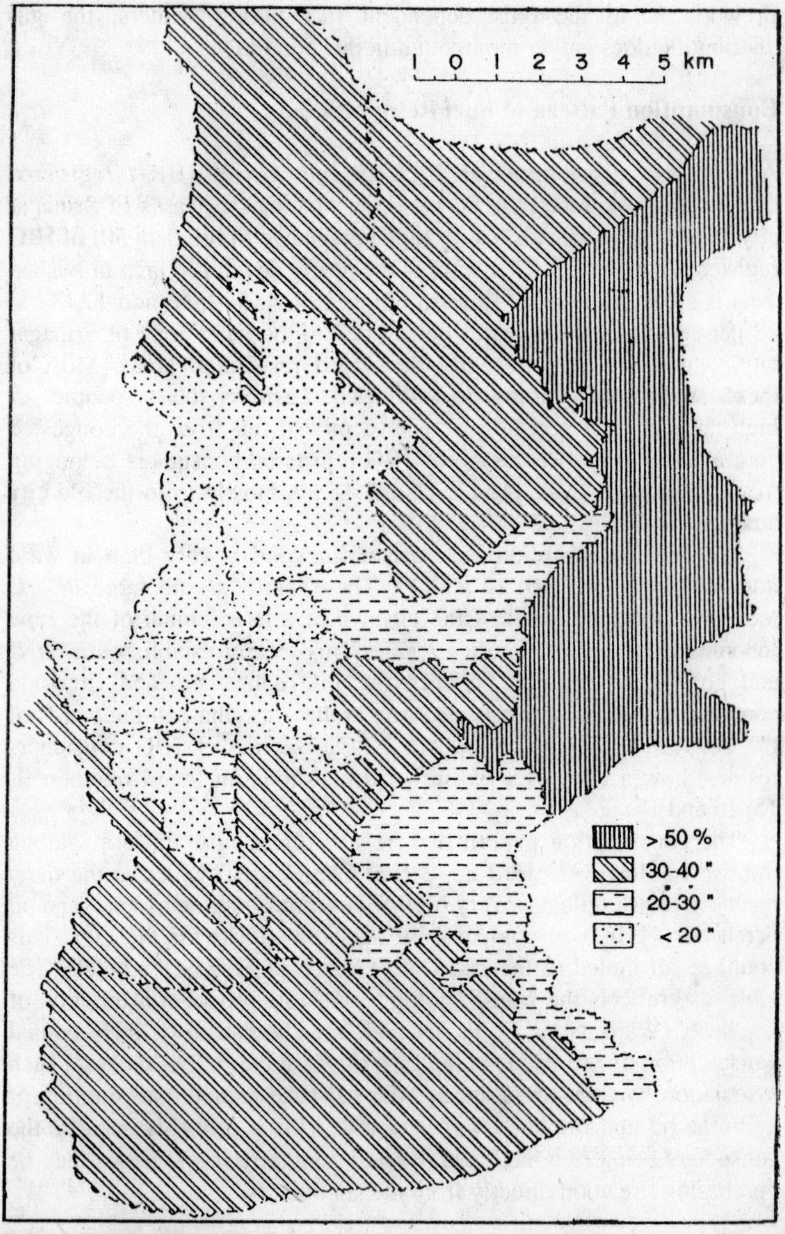

Source: Food & Supplies Department (G.F.) of High Income Group Registered Consumers 198

Fig. 6.3: Srinagar Municipality Percentage of High Income Group Registered Consumers 1984.

TABLE 6.5: **Distribution of Fuel Resources in Srinagar Municipality, 1982-84**

Sl. No.	Months	Firewood (qts)			Kerosene Oil			Gas (No. of Cylinders)	
		1982	1983	1984	1982	1983	1984	1983	1984
1.	December to March(%)	182398 (44)	18032 (45)	178408 (46)	4775 (33)	5567 (35)	5345 (35)	28613 (44)	29396 (44)
2.	April to July(%)	100367 (24)	93981 (23)	82895 (21)	3819 (27)	4450 (28)	3973 (26)	17718 (27)	17981 (27)
3.	August to November(%)	128376 (31)	90222 (32)	130981 (33)	5750 (40)	6048 (38)	6013 (39)	19122 (29)	19472 (29)
	Total % Growth	411141	406904 –1.03	392282 –3.59	14344	16065 +11.99	15333 +4.55	65453	66859 +2.13

TABLE 6.6: **Per-head Annual Consumption of Fuel Energy**

Ward No.	% of H.I.G.* Consumers	Firewood (in kg)	K. Oil (litres)	Gas per 100 popn. (Cylinders)
I	54	41	20	21
II	28	61	35	21
III	31	98	41	38
IV	36	56	35	10
V	34	56	30	7
VI	19	100	25	7
VII	27	111	18	12
VIII	14	111	20	38
IX	30	73	22	38
X	19	112	18	7
XI	17	53	29	7
XII	23	109	30	4
XIII	12	91	18	6
XIV	13	75	26	6
XV	15	37	28	46
XVI	38	38	26	4
XVII	30	39	33	4
Total		68	26	12

* High Income Group consumers are those whose family income is more than Rs. 600 per month.

Source: Directorate of Food and Supplies, Srinagar.

The consumer gas shows a wide range of inter-ward variation. The maximum consumption is in ward numbers 3 and 4 belonging to the areas around Lal Chowk and Nai Sarak.

Hypotheses Tested

The present study has tested the hypotheses stated earlier with the following results.

The correlation coefficient between the percentage of High Income Group consumer to the total population and the annual consumption of firewood per head works out to be –.4640. The *t* test of significance showed a value of 2.45 which is greater than its tabulated value at 5 per cent level of significance. This indicates that the null hypothesis is rejected at 5 per cent level of significance, thereby accepting our second hypothesis that there exists a negative correlation between the above two variables.

The relationship between the annual consumption of kerosene oil and firewood per head shows a negative correlation with a value of

.3945. The t value was found to be 1.96. This was greater than its tabulated value at 10 per cent level of significance. So we may accept our third hypothesis rejecting the null hypothesis.

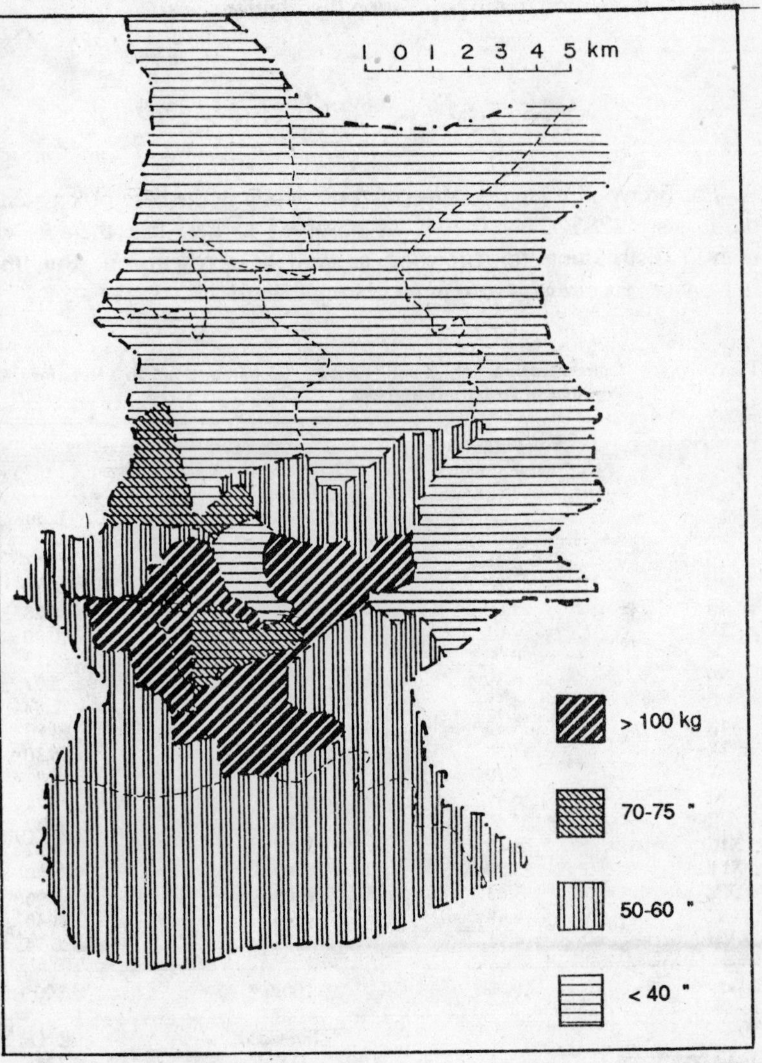

Source: Forest Department Srinagar (G.F.)

Fig. 6.4: Srinagar Municipality Annual Consumption of Firewood per head 1984.

Imbalances in the Energy Resource Distribution

The imbalances can be best studied through Gini's coefficient, which was worked out by the formula for the consumption pattern of energy resources in relation to the population distribution.

$$G = \frac{1}{100 \times 100} | \Sigma_{i=1}^{n} \times i\,yi + 1 - (\Sigma^{n} \times i + 1\,yi) |$$

For firewood it gave a value of .1187 while in the case of kerosene oil it was .1285. This shows, as observed earlier, that there is an uneven distribution for firewood as well as kerosene oil. But, the imbalances for firewood are more glaring than for oil.

TABLE 6.7: **Concentration of Fuel Energy Consumption in Relation to Population Distribution, 1984**

Ward No.	% of popn. to total population	% Distribution of Energy to total distribution	
		F/wood	K.oil
I	5.16	3.11	3.96
II	3.30	2.98	4.40
III	4.00	5.78	6.28
IV	7.81	6.41	10.20
V	8.09	6.61	9.19
VI	6.39	9.46	5.97
VII	4.18	6.91	2.84
VIII	5.12	8.45	3.87
IX	7.58	8.20	6.30
X	7.70	12.47	5.31
XI	6.47	5.11	7.11
XII	7.45	6.65	8.47
XIII	6.35	3.85	4.32
XIV	5.29	5.89	5.20
XV	5.63	3.09	5.96
XVI	4.48	2.49	4.40
XVII	4.91	2.35	6.14
Total	100%	100%	100%
		Firewood	K. Oil
Gini's coefficient		.1187	.1285

Conclusion

The study on the consumption pattern of fuel resources in Srinagar Municipality leads to the following conclusions:

(i) Srinagar Municipality consumes the maximum share of energy resources in terms of kerosene oil and consumer gas in the Valley. This is reflected by the fact that 67 per cent of kerosene oil and 82 per cent of consumer gas is concentrated in Srinagar Municipality alone while the city accounts for only 19 per cent of the total population,

(ii) Registered consumers of kerosene oil are almost uniformly distributed in the wards of the city. In the case of firewood, most of the consumers are concentrated in the old city. This indicates the popular use of firewood among the families located in the core areas. The families residing in the newly settled wards prefer to use the other alternative energy sources. Due to the complicated procedure for obtaining a gas connection, the supply of consumer gas is mostly restricted to the areas adjacent to the main business centre of Lal Chowk and in a few peripheral areas, where high-income families are residing.

(iii) The pressure of population per depot for various energy resources indicates an identical threshold for kerosene oil among the wards excepting the newly developed ones. The distribution of firewood depots showed large-scale variations,

(iv) The percentage increase in the distribution of kerosene oil and gas did not show any substantial decrease in the consumption of firewood. Firewood consumption is very high during December to March due to limited supply of kerosene oil as well as persistent use of firewood for multipurpose functions like heating, etc.,

(v) Firewood consumption is also high in the wards having a large share of low-income group population,

(vi) There exists a negative correlation between the consumption of kerosene oil per head and the consumption of firewood per head annually.

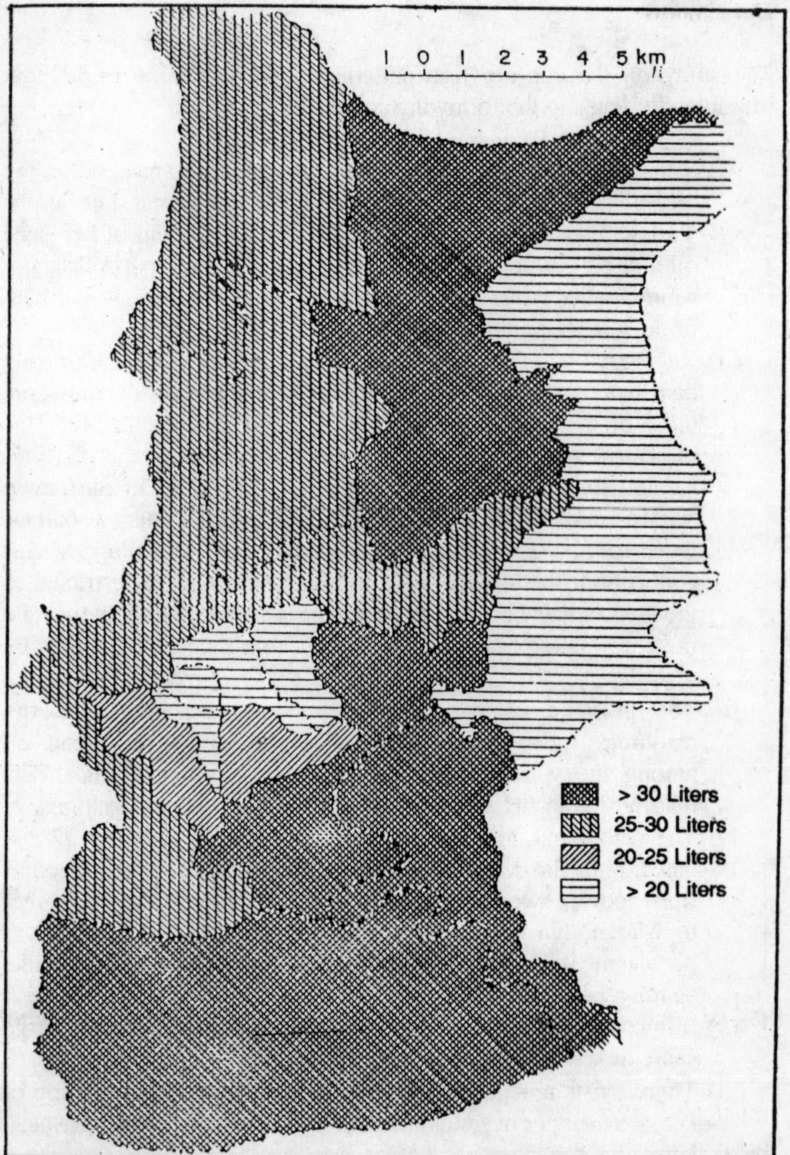

Source: Food & Supplies Department, Srinagar (GF)

Fig. 6.5: Srinagar Municipality Annual Consumption of Kerosene Oil per Head 1984.

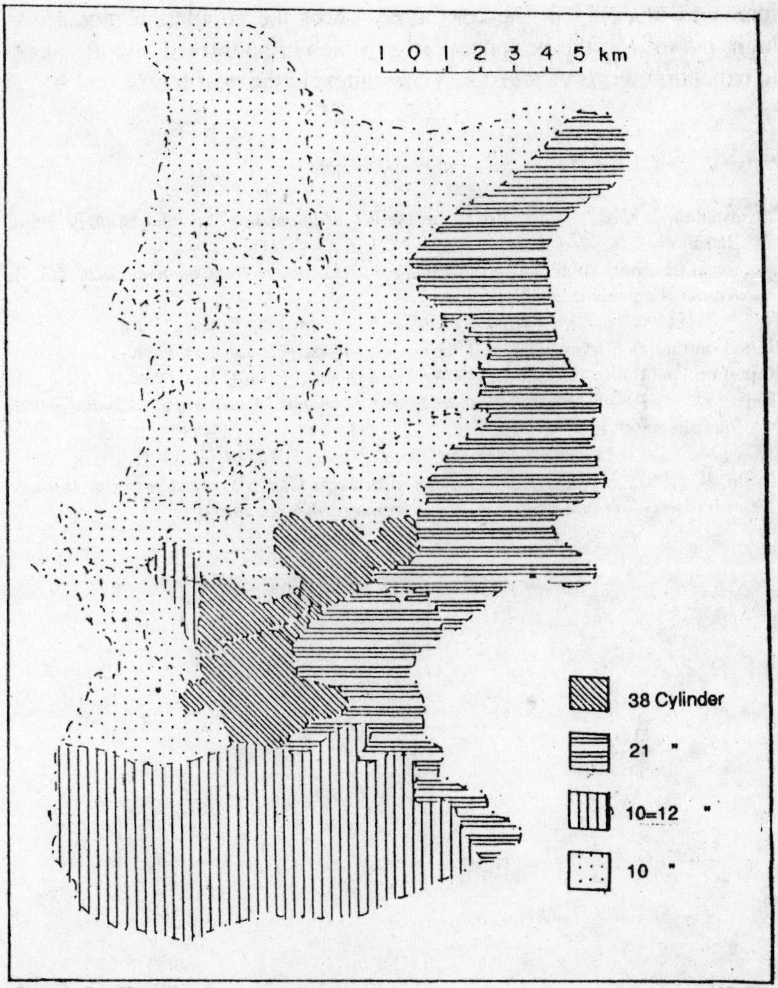

Source: Jay Kay Gas Company (Srinagar) (G.F.)

Fig. 6.6: Srinagar Municipality Annual Consumption of L.P. Gas per 100 Population 1984.

The study therefore suggests that the consumption of firewood must be restricted to the minimum possible extent. This could be achieved by providing the various alternative resources of energy especially that of kerosene oil and gas supply. The shortage during winter may be overcome by creating a buffer stock of kerosene oil during autumn. The number of depots for kerosene oil and gas must be increased not only in the core areas where the population pressure is high, but also in the peripheral areas which are unserved and are likely to expand in terms of registered consumers in the near future.

REFERENCES

Darmstadter, J. *et.al*, (1971) : *Energy in the World Economy*, The John Hopkins Press, Baltimore.

Nagarajan, R. and Lahiri, A. (1977) : *India in Perspective, Development Issue*, Vol. II, Arnold Heinemann, pp. 211.

N.C.A.E.R. (1959) : *Domestic Fuels in India*, Asia Publishing House.

Report on Energy Survey of India (1965) : *Government of India*, New Delhi.

Report on Fuel Policy Committee (1974) : *Government of India*, New Delhi.

Tyner, W.E., (1978) : *Energy Resources and Economic Development in India*, Allied Publishers Pvt. Ltd.

United Nations (1973) : *World Energy Supplies*, Series J, No. 18, pp. 13-35.

Zutshi, B. (1987) : "Energy Consumption in Srinagar City", *Transaction of the Institute of Indian Geographers*, Vol. 9, No. 1, January 1987, pp. 59-69.

SESSION II

CONVENTIONAL SOURCES OF ENERGY IN J & K STATE

Chairman: Prof. Shah Manzoor Alam, Vice Chancellor,
University of Kashmir

SEVEN

CONVENTIONAL SOURCES OF ENERGY

S. VARADARAJAN

Mr. Chairman and Participants

I have still some hesitation in accepting the position offered of delivering the keynote address on conventional sources of energy in Jammu & Kashmir State. It is a subject on which my ignorance is unlimited and it has been only heightened by reading all the papers that have been included in the docket. I would like to first of all commend all those people who have submitted these papers. I think these represent very good collection of data, very perceptive examination of future needs, very thorough analysis of different types of energy that could be developed within the State and what impressed me most was that these were based not on global figures from the World Bank or some futurology studies but on very good studies made locally. For instance there is a very good paper on the municipality of Srinagar although it says it very pampered population in terms of energy compared to all others. There are very good studies on cluster of villages and very good understanding of the needs in different seasons and about variations. I have also seen papers from the Survey of India which outline the serious problems and I am very happy that the university, the departments in the university, scientists from different departments not only from the Science Faculty but from other faculties like Economics, Commerce have all contributed very important papers. Whatever I am going to say have to be really related to this basic document and I hope there will be opportunities later today of analysing the excellent data presented and coming to some conclusion on the future.

As has been said time and again that energy, of course, represents the most important input in the whole way man develops, and the civilization of man is evolved. In the absence of some other forms of energy man tends to use only human energy or animal energy, very rarely anything else, may be some wind energy for sailing. So civilizations have been marked by the extent of which conquests have been obtained over other men using slaves or while we may not use the word slavery in the sense the people are subjected to total control and there are still forms of slavery in the sense that one is required to perform very arduous physical tasks under very difficult conditions for earning livelihood to meet minimum needs. So there can be very few places where this form of human energy inputs are demanding as much as in J & K and over territory given the conditions of weather, the terrain and slopes. It is even difficult to use animal energy in many cases and all one has to do to go little bit outside the city of Srinagar then one realises very large amounts of human energy and human time are used for this purpose. Now the requirements of energy are often, it is said in the papers themselves underestimated time and again. It is said that it is because the Planning Commission does not want to spare money and purposely underestimate. One of the papers suggested that they are not very happy to give larger estimates and so their estimates are conventional increases over the past 10 years and you average them. I think the rate of change of energy requirements is not going to follow that path. I would like to say for many reasons. One is if you take India where we are very proud of very large increases of energy in many ways you find that our energy consumption per capita are still comparatively low. For instance we only consumed one million tonnes of oil or less in 1951 and today some 40 million tonnes not only oil but also gas is insufficient for meeting our needs. We are now estimating that it may go to 95 million tonnes of oil besides gas by another 14-15 years and that may even be an underestimate. In spite of the fact we will have large amounts of gas which are not quite accounted in the system. So one of our worries is that despite the discovery of oil and I think the phenomenal amount of work done for processing it to meet our kind of requirements because the oil that we discovered is not similar to the oil of Middle East etc. Nevertheless our imports of petroleum products will be higher in 1990 than in 1985. Even if we discover some more we will not be ready enough to use for exploitation. Unless we discover other sources and exploit them our requirements will go up very substantially and during the next 10 years it is possible that it will not present too serious a problem because all the indications are that in the immediate future, i.e., next five to seven years there will be drop in oil prices. These may come down to 20 or perhaps 15$ per barrel. There are struggles going on but it is inevitable that oil prices will come down. Now that has some problem because these temporary drops will tend to hide the fact that in the longer term these are not going to be available and that may even alter our investment estimates and proposals whether it is nuclear or hydro or any other. Because these calculations will show these are much less economic. Although there is very large amount of gas in the world which is not yet fully utilized and I believe last week Mr. Gorbachev has ordered very large increase in the use of gas in the Soviet Union so that the U.S.S.R. will be replacing oil so that the oil could be sold in the free market. The Saudi Arabia which produces as much as 9-1/2 million barrels per day at one time in 1972-73 – in order to keep up the prices of OPAEC has reduced its production to three million barrels per day and they have now declared they cannot go below for the reason they themselves consume at least 1 million barrel a day. They need to keep up this average because there are very technical difficulties and they cannot manage their economy with much less.

Now most people think that energy consumption conservation strategy should be to reduce the oil energy consumption by 18 per cent in Europe. So I think the rest of world will also do this with the result that we will have some oil surplus in the world for the time being. That should not hide the fact that our own requirements will steadily increase for a number of reasons. One is population factor. Our population is increasing at the rate of 15 million per year which would amount to an increase of 225 millions in the next 15 years. I think that would be an underestimate as we know 40 per cent of our population is below the age of 16 and there are very clear indications that we shall be able to do very much better about infant mortality and women mortality in coming years. We have abolished small pox and we have abolished cholera but we have very large number of deaths due to other viral diseases. We have large loses of infant mortality and we are among the highest in infant mortality. But I think this will change because of our ability to do something about communicable diseases, amobic dysentery, biotechnology and various things. So our population growth will continue. Therefore, the pressure on energy sources will increases much more than now. The second point is that despite the fact that we are anxious to preserve the rural sector and give advantages for living in the rural I think there will be inevitable increase in the urban population or even semi-urban population. As soon as we live even little bit away from the rural sector there are many changes that take place. Firstly, all people are not employed. In the rural sector even children work in the productive sense (though they are not being paid). But they are doing something productive which is not done as soon as they move to the urban area. Older people are also not useful so with the result many of them will be deriving benefits of one type or other and their time will not be used. I will come to that in a little while. That means they are not available for collecting firewood, bringing material, doing work which is now being done in the rural areas – bringing cow dung or whatever other forms of energy. Large proportion of time of people is spent on these things. Whereas now they have become dependent on energy inputs whether it is for heating, cooking or transport or movement with the result we will be requiring energy inputs in urban areas and it is difficult to see despite of all that very large sums of money going into Integrated Rural Development Programmes (IRDP) which is consuming very large amount of our plan money. This cannot be fully sustained. Some movements will continue to take place as in Bombay or Calcutta or Delhi. They are only indicative of other urbanization processes taking place in districts and so forth. There is no town where the estate prices are not increasing, population densities are not increasing and also the demand for power. Somebody said about Srinagar being pampered which is correct. All the kerosene and biogas is distributed in Srinagar in very large proportion compared to any other region. Uncertainties in urban areas in health and other things are reduced to much greater extent. This will put very large pressure on our requirements.

We also know that we have plan to increase our coal production. We have now 150 million tonnes to 160 million tonnes and we have 30 million tonnes in pit-heads. Our plan is go to 400 million tonnes by the year 2000, i.e., about 220 million tonnes in another 15 years or so. There seems to be determined attempt to improve coal mining and funds are being allocated, not sufficient but good enough to be able to do that, but insufficient funds to be able to move the coal. So the situation as far as the railways are concerned will aggravate during the next 5 years or so, because if we could not move the additional would result in coal that non-utilisation of power plants due to the non-availability of coal or because of poor quality– if transportation lines are extended this will become a serious problem. I think the problem is that railways stock, because they have been using accounting norms which do not include inflation accounting and sufficient depreciation. The result that the stock is pretty old and they do not have money for replacement. I think that just as in the case of our roads the opportunity to increase speed or have much more haulage tonnage it will be more difficult than before. So because of resource-constraints-planning–process this movement of energy systems is going to be very serious problem. When we come to oil I think movement to north is going to be very difficult for the reason that it was planned to have a refinery in the north of 6 million tonnes and in addition to the one in Bangalore. Now both of them have not started. I think it is very doubtful whether we will have any start in the present plan. It will mean importing with some exchange of our crude but we will have to move hydrocarbons from the coasts to the north. The products can not be moved by pipe line. We have no pipe line for products. We have only pipe line for crude. So as soon as we have products pipe line the expenses will be very large because we need separate pipe lines for different products. So I think this is again very serious constraint and the availability of petroleum product towards the end of railway line is always more difficult from the port. I think some study is required how we are going to move for instance if you are saying that there will be increase in the requirements of petroleum products in J & K how are they going to move this and wherefrom and from which refinery? I think the nearest refinery is at Mathura and that is already more than fully committed. There is no other and the next one is Baroni. I don't think anything will come from Baroni. We are doing big job of further revamping and debottlenecking the refineries but in the end you will have to import about 10 million tonnes of petroleum products, kerosene, light diesel and so on. Which will mean they will move from Kandla, Madras, Bombay and these are already fully loaded. So I think there will be problems of this nature as well. Now the solution that is being offered, I notice consistently is to assume that the movement of coal from outside the state is fraught with so much difficulty that you cannot make a serious account, take serious note of it. I do not know whether it is correct or not. All the papers I have read which have been put up here state that the opportunity for the movement of coal into the state to any large extent is very limited and so we have to find out other means. What are these other means?

Today's seminar will obviously attend to it. All I can say is that the only thing being offered is fraught with very serious difficulties. If you look at the next 5 years – may be if you have an integrated plan for the 10-15 years you may be able to overcome many of the problems. Because in 5 years and knowing the history of previous budgets I do not know whether we can expect very large increase in hydel power generation. Each one of this I presume that the ones you will take up are all going to be in terrains more difficult, less accessible with more problems than the ones which have been tackled. So to that extent much integrated planning is necessary and solutions will have to be found and there is no suggestion that such a planning and execution process mechanism already exists in this state. Although it is stated that hydropotential, if you take up everything, i.e., total glacier, total melt capacity, has 100,000 MW. But you could easily generate 10,000 MW installed capacity. I think the proposals that are all put up are not adding up to it except in one or two cases even to 100 MW. I think it will be very useful to have an open debate on the manner of execution of hydroprojects in the state in this part. Considering the time factor for the setting up of 2-1/2 – 3 MW we can make projection based on the current inputs, organisation and management of the time that would take to add to the additional capacity for this purpose. So that is an area which is most promising from all the papers that I have read and most compelling. I believe the alternative to that will be to import large quantities of petrol.

One has to have some method of stocking operations of energy materials according to certain minimum needs which we are doing. For instance, we have stockpiling of petroleum crude because of uncertainties of imports. We have put in very large amount of crude storage. We have had a plan of 90 days but I think people are pressing for 180 days. This is a very large amount of storage which is required, so some of the storage of petroleum products has to move and some capacity has to be established and so has been established with the oil companies individually. But there is very large outfit in Delhi for the petroleum distribution and stocking organisation and it is very easy to get figures from them and see how best this stocking can be done.

It does not mean that we could not at all have some amount of stocking of coal and since you are moving through difficult terrain and using railway systems which are loaded one must move such coal which has high energy content. Now that might be difficult because we have to make a distinction between high ash coal and low energy coal and so on. It is difficult but nevertheless if you are to spend money for moving something, why move something which is unwanted. And, therefore, better quality materials will be appropriate for this purpose.

It is also suggested that there are other conventional sources. The whole of India still uses 50 per cent of its energy requirements from renewable sources, mainly firewood, some bits of hydro, very small. Still now the problem is that as our development takes place including agriculture, I do not see how we can sustain this proportion merely because few of us have got some new initiatives in non-conventional energy sources. We talk a lot about biogas. We talk a lot about firewood and biomass. There are many scientific organisations working on it. One, it does not mean they will really make very substantial contribution at a time when there is very sharp rate of rise in the use of energy. I will really give you some comparison. We use some 40 odd million tonnes of petroleum products for about seven hundred and forty-five million people. The United States which has about two hundred and thirty odd million people uses about 900 million tonnes of oil and about 600 to 700 million tonnes of oil equivalent of gas. We have 160 million tonnes of coal and our usage is less than that for this population which is three times the U.S. population. The United States is close to thousand million tonnes of coal of somewhat higher energy content than ours.

Now, the situation in the USSR is the same. Or, you take China. Of course, climatic conditions demand slightly more but nevertheless the order of magnitude is very different. China today, this year, pushed up the target of coal production even further. It is reported to increase to 700 million tonnes for a population which is not unlike ours. Yet we are talking about their population. It may not reach, a billion in the year 2000. They are in excess of ours presently and yet we are only talking about a target of 400 million tonnes of coke against the Chinese present target of 700 million tonnes.

Probably they will exceed a billion tonne. The reason I mention is that even when we congratulate ourselves on our food production, we must look at numbers. Our food production this year was 152 million tonnes and we said, we cannot consume all of it and we may have a stock of 25 to 28 million tonnes, may be surplus in an overall context after consumption of the order of 8 to 10 million tonnes or even export for a population of 750 million. The United States with a population of 230 million produces 250 million tonnes of grain and has some surplus of 15 to 20 million tonnes. Russians again have a similar population.

They produced 230 million tonnes and they are short of 30 million tonnes of grain and they are buying from us also. They import 20 million tonnes from the United States, that is, the requirement of human energy as people get more ambitious and more developed is very substantial i.e. it is three or four times even in grain consumption for a number of reasons – one is the levels of consumption are higher also many of them go to meet economy which requires two and a half times of input energy to be converted into meat but on the other hand, our agriculture today is going to require very large energy inputs and we have not made a proper calculation. I will give you some examples – we used only hundred thousand tonnes of Nitrogen in 1951. We produced about 52 million tonnes of foodgrains. Today our consumption of fertilizers. . . . Nitrogen is about 8 to 9 million tonnes. Out estimate for the year 2000 when we want to increase our grain production from present 152 to about 240 million tonnes which is not too ambitious considering, we will be adding 225 million people which is one-third as much so and there will be some wastage as urbanisation takes place. We will require an input of fertilizers of the order of 19 million tonnes. Again fertilizer is made from energy. We use hydrocarbons to produce ammonia. We remove Oxygen from water to remove it as carbon dioxide. We let off carbon dioxide and fix ammonia as Urda with carbon dioxide. So this, we for some carbon source so we will require a very large increase in carbon sources for production of just pesticides. It is estimated that our requirements of pesticides may be as high as 60,000 tonnes even export in India. We are very low in pesticides that because we are going to reach the targets of good production, we cannot do it unless we go to more intensive agriculture. Our production of grains is less than two tonnes per hectare whereas we can demonstrate six and a half tonnes per hectare in any agricultural form or agricultural university and the last year U.K. with much less favourable climate produced an average of 9.3 tonnes per hectare and peak ones showing up to fifteen i.e. we have opportunities without increasing land acreage or even water to be able to produce most but it would require input energy, fertilizer, pesticides and also site developments and perhaps use of plastics so and so forth for conservation of water, energy of various kinds and the reason I mention is, it is calculated that in the United States every calorie of grain produced requires input of 2.3 calories of non-conventional material, hydro-carbons of one sort in order to produce one. Now that is not so tragic except when it comes to consumption because of the economy and the most efficient is chicken which can produce about 2.8 calories per calorie of meat produced. Therefore, it is estimated that it is about 7.5 or so calories when food is actually produced and by the time it is estimated that it is canned and eaten for every calorie consumed by human being in U.S. and average ten calories of input of non-conventional, non-renewable energy is put in. I mean it is very stupid to do so now we have not really worked out a plan how we shall increase our food production by methods totally different from those in the developed world. The developed world has in fact used the cheap energy of pre 1973 as major input for agricultural production and they all have, I don't know, 10,000 acres farms or 5000 acres highly mechanized but very high yields. Now here, I think, we have no strategy as far as I know, may be you can ask our agricultural scientists we should give us increased yields without not proportionate disproportionate increase in the use of energy in the farm.

People may not realize it because you say we are still tilling. We have got small farms. We don't have big diesel combine harvester and fertilizer and all that but as out production increases intensification of use of fertiliser and big complaint is that our use of fertilizer is still too low in many States and pesticides is very low indeed and all this we will have energy conservation will also require water conservation. We use about roughly three times water as much as Japan does per each tonne of rice produced. We can produce more with the same water provided you can store it. Provided you can channelize it in appropriate amounts, preventive operation and so on, all of which will require new materials – cement, plastics, pipe lines etc. So I believe agriculture is going to demand very much more. I don't know anything about the input of materials into the agriculture in J&K and there is no paper on it. I presume that the inputs are relatively low, fertilizer, pesticides and other things are not high if the pressure for increase of production in agriculture is to come and that will come necessarily because I believe you are subsidising food in a very substantial way much more than what is done in other States than this subsidy will become very large if something is not done for productivity increase so of productivity is to be increased then energy inputs or energy materials whether it is ammonia or fertilizer or some thing more will have to increase very substantially and that is a matter which has to elaborate and energy today is being looked at only in terms of direct energy inputs in all the papers for mechanical work, for heating so on but no estimate has been made of energy that goes in the secondary processes or energy based materials which have to come into the States which are not made here. There is no ammonia made here. So some estimates of these have to be made in relation to agriculture and if later on you are going to have energy input materials like cement, concrete, plastics etc. All that is required for water conservation and this order of hydro power and combine methods cannot be done unless very large amount of inputs of building material or containing materials, container and caps what else are to come in and of this food preservation and so is going to increase there is going very substantial requirement of energy for refrigeration. It is curious in the State which is cold for many months. You would also require in other months for food preservation very substantial order of these are high value products, fruits, fish or meat or what have you all of them are high value products and either you can dry them or you can freeze them so you will have to use any process.

Now we come to see the opportunities much of what has been suggested here, say that we can that 50 per cent required is met by fire wood. I believe it is wrong to look at energy in terms of fire wood as equivalent to say kerosene for simple reason that kerosene is delivered energy and you just use it. Light it and use but in the production of firewood in the farming or in forestry and then it needs felling of trees, cutting them and then you have to dry the material in sun-light or some times in kilns of various kinds and then it has to be transported to the points of consumption by not the methods by which you deliver petroleum products which is in tankers or rivers so on so forth but by human energy and sometimes animal energy. I think we have to see just as we see in agriculture the net energy that is available when you have one kilo of firewood for getting one kilo of firewood, you are putting a lot of human energy forestrying transportation and in sun light what have you and I believe that is going to be limitation for the production of firewood. I think this assumption is made merely because you can increase biomass. You will produce very large amount of firewood is not enough because in all other parts of the world, there is firewood available. In fact they even set fire to forests. It is not unknown even in Assam where if you go to the north east where from time to time people set off fire to forests.

In order to see that the forest does not encroach urban land and you can see big fires there. This was common here. I presume some fifty years ago or so and so on, it was not even uncommon in South India where I was brought up, to set fire to forests so that it does not encroach your land.

Now on this some study has to be made. It is absurd for me to suggest it is large or small but if you say the amount of labour cost is going to increase today, for instance, in Punjab, it is about Rs. 65 a day during harvest time at least and it may be even in south about Rs. 30 or Rs. 40 a day so how are you going to get enough people to be able to do this physical work in order to bring whether it is charcoal or firewood and so on so forth to the point of consumption in a quantity that is required. Then I would also like to pose another question whatever happens in the rest of world should not affect us but in a democratic free world it affects us we are less proven and less inclined to use materials which require a lot of storage, lot of effort in using than something which is readily available so every body wants gas, a cylinder of gas. Oh! we can say a cylinder of gas is much more expensive than kerosene; it is more expensive than wood or coal or charcoal or even smokeless fuel which is now made and available but people wouldn't use it because their transformation in standard of living is taking place and therefore electricity and readily usable forms of energy which do not require physical storage of this quantity is now preferred whether it is in Bombay even the poor has preferred it and they will not make this calculation saying it is cheaper to do it and you can spare that money for something else say for food. Now, I think by time you move into urban areas, you see very big constraints and the value of people. Time is going to be very big constraint in the use of alternative forms of energy. Now you can say, you are going to take four hours to cook what does it matter. I think even in villages getting four hours for your solar cooker to turn out your rice is not easy and some body has to sit around, to see that the dog doesn't eat it and so on and I mean these are practical problems. I believe only people. I have heard people in Delhi praise it but I believe there are only people who can have a terrace or something and saw people cook these. I do not know. I think that some study has to be made of this sort of a thing to see whether this is practicable. The reason I mention is that if you do not have the studies and if you make the assumptions that investment in forestry biomass and so on will fulfil the requirement and find that there are other constraints of various nature as has happened in the western world or anywhere else. Then we would have invested in transport or other mechanism and would have found ourselves in great difficulty.

I believe one item which requires greatest attention is energy conservation. You know on the whole we are all used to in this country a feeling of plentifulness whether it is water as long as we throw any thing into the moving river it gets oxidized and in Ganga or Jamuna perhaps everything is very pure. But there is limit to which organic material can be handed to water resources, exceed their oxidation capacities and we have now dead rivers like the Damodar or Bukharo. These are good names but they are all dead i.e. they have gone beyond the point where any sustaining chain of life can be made at all. It is impossible so photosynthesis doesn't take place at all so I think we here also have to see the problems of pollution and so on that will come by increase in use of energy which lead to it and we have to get away from the idea that every thing is plentiful like water or energy or firewood. So we will have to now look at it how to consume energy and make capital resources for this purpose cheap enough so that conservation is possible. I think the real problem of today is the capital cost of installing materials for conservation are sufficiently high and funds are always in short for any body not only a State capital is short but an individual family capital is short. So consumation is not possible unless you can amortize it for a period of ten months or twelve months and have that amount delivered to you preferably not as cash but in some kind then it is possible to consume energy in an efficient manner. I think this is the most important input that the State has to make almost all devices that are used here are much worth of it. There is much greater need, for instance, in the Transport sector for conservation let us say in Bombay in our pricing policies we do not take into account the average cost of transportation of all petroleum product. It is a real cost and so it will be very helpful if some Government does something to improve the operation of lorries and especially in this State because that will reduce the consumption which would give enormous advantage. It may not at all be pricing policy but nevertheless there is very good reason for producing efficient high energy as a consequence one must go to the point of saying the most efficient energy devices should be first demonstrated here not necessarily in the places very near us because you have such a long line and it would be great advantage to demonstrate it here and like all things unless you demonstrate something extraordinary. It does not filter down to the ordinary so by doing something quite extraordinary, I do not know, seventy, eighty miles per gallon vehicles which are round the corner in Japan now they have already selling 70 miles per gallon to U.K. and so on. I do not think, I make some such things. If you demonstrate it, than 18 miles or 20 miles to the gallon will look so poor that at least people will try to improve it to 40 miles per gallon. I mean this sort of things of excellence in the demonstration will convince the people a high nature which require high capital but improvement will be one way of drawing attention to these things people do not believe you, I think none of you believe me when I saw it comes to the gallon even now even though I say it but it is true 70 is advertised guarantee at 56.6 miles per gallon is the standard U.K. test for an automobile far a passenger care. So I believe these are all something which we have to look at. The second is that I believe there is no State science State Council of Science and Technology here.

There must be some place where applied knowledge can produce enormous benefits. I find that very little things that we do through our CSIR laboratories have a tremendous effect and these produce all kinds of changes in employment pattern, energy usage and all that and which do not happen so easily say in Bombay. No body takes notice of it. They get all their input from London, Paris and so on, now this year every thing that we do get used and I believe the State Council of Science and Technology, if it is formed with some input for transfer of technology demonstration risk founding will all be very good. There would be a need for some form of State finance corporation specifically for high risk energy projects because the ordinary financing systems will not be able to demonstrate these things and we have to take some risk funding you may get it from the Centre. Mr. Dayal is coming, you can ask him, he is also proposing a new bank but I think the State itself will find it highly profitable to take high risk projects and demonstrate them if it not from a pool of funds but from a source which is earmarked for this purpose. I believe the state planning processes will require much sharper quantitative estimates of energy not in isolated system and people say – I can preserve this food for this but one does not calculate all the other energy inputs which go into that. So you will have to make much more technological system studies than have been done possibly in the past and if that is done the options will be little better. I think I am not saying that any mistakes are being made but getting everything into one unit i.e. the energy units are now used commonly whether it is in mechanical system or any other common unit in language that is used will be very useful. All this cannot be done inspite of the very good papers in a sustained basis by holding seminars. One has to have an organisation for this purpose. In a state like this there is no gain saying the fact that the key input which will determine development and the cost and its future is energy. All else is subsidiary to this because the requirements relative to other people in other parts of the world are large. The availability of energy is quite small.

I can talk about Geothermal energy. A survey made proposing projects making one megawatt generator by using steam available at 125-135C° and so on but is not gone up very much beyond that. May be we should do those things. There is some paper on wind energy and wind data collection but they wouldn't contribute much unless we invest substantial amounts and have wind farms say for 20-25 acres. So I think we need a perspective plan very much more than any where else and that would require inputs of new things like risk-funding mechanism, a State Council of Science and Technology and energy advisory board for local requirements which will take into account all the inputs and then see what would be the most appropriate action. If as a result of this seminar some initiatives are take for this purpose, it will be your problem because your State Govt. does not have a separate energy department or a board. It has many different things and one cannot take an overall view of these things.

I am afraid I know very little about J&K. All I have is only these papers. I can only guess what might be required. I hope this will be of some use to you.

Thank you,

EIGHT

CONVENTIONAL SOURCE OF ENERGY IN J & K STATE

B.N. KAUL

INTRODUCTION

The State of Jammu and Kashmir is the uppermost riparian State of the Indus river system. Its area is 2.22 lakh sq. kms. including those areas under occupation of Pakistan and China. Its population as per the 1981 Census is 59 lakhs. The net area of land sown is about 5.15 lakh hectares. The forest area is about 22,600 sq. kms. which forms about 8 per cent of the total geographical area. The remaining area is mostly mountainous lying in the rainfall shadow of the cold desert of Ladakh and Kargil Districts.

Topographically the J&K State can be divided into three distinct regions. The Jammu region consists mostly of mildly rising hill slopes from the plains of Punjab to the sub-Himalayan ranges (known as Shiwaliks), having an average altitude of about 600 metres. On the northern side of the Shiwaliks, Chenab river generally running along the strike between Pir Panchal and Zanskar mountains cuts in a very deep gorge. It drains about 8,200 sq. miles of catchment area near Akhnoor.

The Kashmir Valley is a saucer-shaped valley in the synchlical basin lying between Pir Panchal (on its south) and off-shoot of Zanskar range (on the north).

The various valleys in Leh Division fall in between ranges of:

(a) Ladakh,
(b) Zanskar, and
(c) Karakoram.

The average level of such valleys varies between 3,000 metres and 4,000 metres.

Climatology of the State

From the climatology point of view also this State has three distinct regions of Jammu, Kashmir and Ladakh.

Jammu Division has a climate more or less similar to that of the plains of Punjab with the summer temperature as high as 48°C and the mean temperature of 24°C. In winter, the maximum and minimum temperatures range between 20°C and 30°C. This area has plenty of monsoon rains from the south-east. About 800 mm of rainfall occurs in the monsoon. Rainfall in the non-monsoon period is nearly 500 mm.

The Kashmir valley has a temperature climate with a maximum temperature, in summer, nearly 35°C and a minimum mean temperature of 20°C. The temperature in winter may fall to sub-zero. The major portion of the rainfall occurs due to western disturbances in spring. Out of 700 mm rainfall, only 200 mm occurs in the monsoon period. The major portion of the non-monsoon precipitation occurs in the form of snow which acts as a water reservoir for the summer flows.

The areas of Leh and Kargil in Ladakh have extreme climate with the mean summer temperature as low as 35°C and hardly 7°C at night. The entire area falls in the rainfall shadow. The annual rainfall is only about 90 mm in Leh and about 230 mm in Kargil. The minimum temperature in winter can be even lower than -30°C.

Water Wealth of Jammu and Kashmir State

In the ultimate analysis it is the rainfall and snow reserves in the form of glaciers which contribute to the water wealth of any country. A statement of month-to-month (normal) rainfall in Jammu, Srinagar and Ladakh is given in Annexure I. It will be seen that the rainfall in Jammu is nearly 1,020 mm, in Srinagar 650 mm and in Ladakh 91 mm.

The rainfall coupled with the snow reserves gives a yield of about 100 lakh aft. in Chenab, 67 lakh aft. in Jhelum, about 65 lakh aft. in Indus and about 9.5 lakh aft. in Kishen Ganga (Annexure-II).

The total annual out-flow of the Indus system at Karachi is reported to be about 168 million aft. Out of this the out-flow of various tributaries coming from J&K State and other parts of India is as under:

Jhelum	=	67 lakh aft.
Beas	=	129 lakh aft.
Sutlej	=	100 lakh aft.
Chenab	=	100 lakh aft.
Ravi	=	64 lakh aft.
Indus	=	70 lakh aft.

Of the above six rivers the total annual water wealth of J & K State can be guesstimated at about 230 lakh aft.

Hidden Energy in Water Resources of Jammu and Kashmir State

The study/identification/exploitation of hydel energy is a continuous and fascinating science to be enriched by ingenuity of the State engineers themselves. Considering the fact that presently we have a long way to go it will be relevant to make a guesstimate of water resources and hydel potential of such resources in J&K. This has been attempted as under:

Indus River

The Indus river has four major tributaries in Leh and Kargil districts. The main Indus river drains the range between Ladakh and Karakoram flowing through Tibet. Its catchment at Karoo is 15,000 sq. miles.

Shyekh River

This river drains the ranges between the Karakoram and the Nanga Parbat and pierces through the greater Himalaya near Askerdoo in Pakistan-occupied Kashmir.

Zanskar River

This river drains the valley between Ladakh and Zanskar and meets the Indus few miles downstream of Leh.

Suru River

This river drains reaches between the greater Himalaya Pir Panchal and Zanskar ranges.

On the basis of the meagre data available for the Indus, it appears that this river has an annual flow of about 13 lakh aft. Although no discharges of Zanskar river have been measured so far, it appears certain that its annual discharges would be more than of the Indus at Leh. Shyekh river has similar discharges. All these rivers fall from a height of about 14,000 ft. to about 8,000 ft. The annual flow of these four rivers is estimated at 70 lakh aft which gives a mean discharge of 9,500 cusecs. Theoretically hidden energy thus works out to 9500 × 6000 = 3000 MW years. How much of this can really be exploited is a matter of continuous study in the years to come.

Water Wealth of Kashmir Valley

The outf-low of river Jhelum draining Kashmir Valley at Baramulla is of the order of 67 lakh aft. In addition to this, river Kishen Ganga which drains the northern slopes of the Dewlatdar mountains and the southern slopes of the Nanga Parbat brings an additional flow of 9.5 lakh aft. Most of this water falls from reaches about 3,500 m to 1,200 m. Jhelum river as such has a theoretical hydel potential of 3,000 MW years. Considerable work on the identification/investigation of various hydel projects up to a total installed capacity of 1,747 MW has been done in Kashmir Valley. Out of this, an installed capacity of 256 MW has been found feasible on major tributaries of the Jhelum such as Sindh, Vishow, Lidder, Rambiara, Sukhnag and Ferozpore Nallah in the immediate future. The list of identified projects and projects under investigation is given in heading Kashmir Division. In addition to these projects the 480 MW Uri Hydel Project which has the techno-economic clearance is awaiting investment decision of P.I.B. through the N.H.P.C.

The Jammu Division which has been sliced by the mighty river Chenab is the richest part of the State so far as hydel energy potential is concerned. Considerable work has been done by Investigation Agencies of C.W.C., Government of India/J&K State. Himachal Pradesh has also done considerable work on some proposed schemes on the Chenab in their territory.

Theoretically this river has a hydel potential of about 7,500 MW years. In other words the three main river systems of J&K State, i.e. Jhelum, Chenab and Indus, in between themselves can be safely guesstimated to have a theoretical hydel potential of about 13,000 MW years.

Hydel Energy is Key to Prosperity

According to a study made by some power engineers 270 MW of electricity at 6 per cent load factor for one year is equivalent to 340 tonnes of diesel oil. On this basis the above theoretical hydel energy would be equivalent to 20 million tonnes of diesel oil. The above figures have been quoted just to give an idea of the richness of J&K State so far as hydel energy is concerned, and this should leave nobody in doubt about the fact that the key to the prosperity of J&K State as well as the neighbouring parts of the State lies in exploitation of this energy to the best possible human capability.

Perspective Plan of Power Development for next Twenty Years

At present Ladakh region comprising Leh and Kargil Districts, having a geographical area of 95,000 sq. miles and a population of hardly 1.3 lakh souls, is separated from Kashmir Valley and the rest of the country by 11,800 ft. Zojila Pass which is snowbound for six months. Unless there is all-the-year-round communication to this area and possibilities of transfer of power become economically viable, the policy of hydel energy development of Ladakh will be confined to meeting the local needs only. Accordingly the emphasis in Leh and Kargil Districts will be on developing power in small hydel stations in the first instance and as and when transfer of power becomes technically and economically viable bigger hydro-electric projects can be taken up Stakna Project (installed capacity 4 MW) in Leh District is nearing completion, while Iqbal Bridge (3.75 MW) in Kargil is under execution, and in the immediate context of the situation, following more mini micro/small hydel projects have been conceived and are likely to be completed within the next few years.

Investigation of other major hydel projects utilizing a fall of Indus river from 14,000 ft. to 8,000 ft. in Ladakh valley and utilizing of fall of Drass river from 12,000 ft. to 8,000 ft. and Shyekh river from 14,000 ft. to 8,000 ft. will be continued simultaneously with construction of the below listed projects.

Small/Mini/Micro Hydel Projects

		Installed Capacity	Energy (million units)
1.	Igo-Phey	5.6 MW	38.65
2.	Mattayan	45 KW	0.26
3.	Shakerchiktan	260 KW	1.56
4.	Tambis Kunur	200 KW	1.20
5.	Haftal	1000 KW	2.172
6.	Kumdock	100 KW	0.60
7.	Dumkher (Micro)	2 x 1000 KW	2.20
8.	Tungste	100 KW	0.60
9.	Bung Dong	200 KW	1.20
10.	Thusgam Barso	200 KW	1.20
11.	Marpecho	200 KW	1.20
12.	Hanu	750 KW	1.50
Medium Hydel Projects			
1.	Panikher-Parkachak	9.5 MW	21.19
2.	Chotak Project	9.5 MW	21.19
3.	Dumkher (Medium)	9.5 MW	21.19

Kashmir Division

Since small and medium hydel stations have a small gestation period, the investigation policy of the Power Development Department has been re-oriented. It has been decided to concentrate on such schemes which are easy to construct and have short gestation period. From this point of view following schemes are being executed and investigated at present in Kashmir Valley (Annexture III).

However, in addition to the above schemes one more major scheme of Kashmir Valley, URI (480 MW) is expected to be taken up for construction.

It is expected that by the end of next 15 years investigation of the following relatively more difficult projects will be fully completed and these projects taken up for execution after 15 years:

1. Lidder Stage-I = 50 MW
2. Lidder Stage-II = 50 MW
3. Sonamarg Storage = 85 MW
4. Ganderbal = 90 MW
5. Kishen Ganga Diversion = 390 MW
6. Vishow = 9 MW

Name of the Scheme	Installed capacity	Name of Nallah
1. Upper Sindh Hydel Project Stage-II	105.0 MW	Sindh nallah
2. Nunwan-Batkote	22.6 MW	Lidder nallah
3. Shitkari-Kulan	84.0 MW	Sindh nallah
4. Boniyar	4.8 MW	Boniyar nallah
5. Hirpora	9.0 MW	Rambiara nallah
6. Athwatoo	7.5 MW	Madhumati nallah
7. Brenwar	2.0 MW	Doodganga nallah
8. Tangmarg	2.6 MW	Ferozpore nallah
9. Mutchil	200 KW	Matchil nallah
10. Asthan nallah	1.0 MW	Gurez nallah
11. Keran	2 x 300 KW	Keran nallah
12. Wangath	1.55 MW	Wangath nallah
13. Ardow Stage-I	7.50 MW	Lidder nallah
14. Ardow Stage-II	3.0 MW	Lidder nallah
15. Laripora	2.5 MW	Lidder nallah
16. Erin	3.0 MW	Erin nallah
17. Pahaloam	3.0 MW	Lidder nallah

Name of Project	Installed Capacity	Head	State where located/ Investigating Agency
1. Gyspa	225 MW	155/350 m	H.P./H.P
2. Bardung	54 MW	60 m	H.P./H.P
3. Seli	87 MW	85 m	H.P./H.P.
4. Roli	295 MW	270 m	H.P./H.P.
5. Kirthai Naunatoo	750 MW	383/307 m	J&K/J&K
6. Naunatoo-Naigad	400 MW	189 m	J&K/J&K
7. Dul Hasti	390 MW	236 m	J&K/CWC
8. Ratlee	170 MW	113 m	J&K/CWC
9. Bagliar	220 MW	135 m	J&K/CWC
10. Sawal Kot	400 MW	196 m	J&K/CWC
11. Bursar	275 MW	365 m	J&K/CWC
12. Pukhul	325 MW	599 m	J&K/CWC
13. Salal	354 MW	96 m	J&K/CWC

Jammu Division

In Jammu Division, following schemes are at various stages of investigation. When all these schemes materialize the entire hydel potential of river Chenab will be utilized:

The total installed capacity for the above schemes comes to 3,945 MW. In addition to the above, the investigation for the following schemes is in progress. These investigations are being carried out by the Investigation Wing of the Power Development Department.

Name of Scheme	Installed Capacity	Nallah/River
1. Lower Kalnai scheme	40.0 MW	Kalnai nallah
2. Parnai Power-cum-Irrigation scheme	44.0 MW	Swaran river in Poonch
3. Abs Stage-II	60.0 MW	Ans River
4. Rajouri	3.0 MW	Rajouri nallah
5. Chenani Stage 2&3	2x4.0 MW	Tawi river
6. Sewa Stage-II	90.0 MW	sewa river
7. Sewa Stage-III	6.0 MW	Sewa river
8. Hanswar	0.6 MW	Paddar nallah
9. Naiged Scheme	4.0 MW	Naigad nallah
10. Bichlary Stage-I	3x5.0 MW	Bichlary nallah
11. Bichlary Stage-II	30.0 MW	Bichlary nallah

It will be seen that the Investigation Wing of the Power Development Department of J&K has already taken up schemes of 54 MW capacity in Leh and Kargil, 726 MW in Kashmir Valley and about 301 MW in Jammu Province under active investigation. There are some more schemes on Kalnai nallah, Ans nallah and Tawi of J&K State which have already been conceptually evolved. Investigation for such a scheme will be taken up after the final project reports of the abovementioned schemes are completed. Many more nallahs, and major tributaries like Mursodher river, Deesa nallah, Gulab Gad nallah, etc., are yet to be explored for exploitation of hydel potential (Annexure-III).

Subject to availability of resources, following hydel projects are proposed to be taken up and completed in the next six years:

1. Upper Singh Hydel Project Stage-II = 100 MW (70 MW Ist phase)
2. Karnah = 2 MW
3. Boniyar = 4 MW
4. Athwathu = 7.5 MW
5. Brenwar = 2 MW
6. Hirpora = 6 MW
7. Pahalgam = 3 MW
8. Chenani Stage II & III = 6 MW
9. Sewa Stage III = 6 MW

In addition to these Salal (345 MW) will be completed. The Dul-Hasti Project of 390 MW, being constructed by N.H.P.C., is also to be completed during the Eighth Plan.

The above projects are expected to be completed partly during the Seventh Plan and partly during the Eighth Plan. Besides the above schemes, the projects to be taken up for execution are (1) Nunwan-Batkote, (2) Shitkari-Kulan, (3) Lower Kalnai, (4) Parnai, and (5) Sewa Stage-II.

The projects, subject to availability of resources, can be completed in the next 10 years. In addition to these projects, Micro/Mini/Small Hydro-electric Projects having a total capacity of 20 MW may be taken up in Ladakh and other parts of Jammu and Kashmir State.

Other Sources of Energy

Simultaneously with exploitation of hydel energy it is also very important from the total development point of view that a huge area of mountain slopes is exploited as far as possible for growth of energy plantation in the form of trees. In this connection, some broad points keeping in view the topography and climatology of these regions could be made and are enumerated as under:

Ladakh and Kargil Area

In this area as discussed there is tremendous aridity and very little soil moisture. There are a number of nallahs draining long valleys fed by glaciers which have surplus water in the four months of summer. Some well-known nallahs are: (1) Wakha nallah, (2) Shanker Chiktan nallah, (3) Vanla nallah, (4) Basgoo nallah, (5) Dumkher nallah, (6) Phoy nallah, and (7) Igo-Mercillong nallah.

In these nallahs most of the summer discharges go waste. Therefore, instead of wasting the summer flows, these should be used for soil moisture conservation measures. The idea is that surplus water of these nallahs instead of going waste as surface drainage should be diverted to the soil and made to seep into it by contour trenching contour bunding, percolation tanks, etc., and where possible small dams.

It is suggested to divert these nallahs into hill slopes, sub-mountain slopes and conserve the moisture as long as possible on these. Suitably designed moisture conservation measures are definitely going to help greenery, growth of shrubs and trees on these mountain slopes and improve ground water availability. In course of time these moisture conservation measures coupled with increasing availability of power for lift irrigation may usher in a green revolution in Leh and Kargil Districts. It is for the soil conservation and forest experts to prepare a comprehensive development plan for these areas after launching some pilot schemes on experimental basis.

Kashmir Valley

Due to high intensity of population in the valley, the pressure on vegetable cover on hill slopes is tremendous. The districtwise barren areas in Kashmir Valley are as under:

Anantnagh	=	17000	hectares
Pulwama	=	10000	"
Srinagar	=	8000	"
Badgam	=	14000	"
Kupwara	=	15000	"
Baramullah	=	22000	"

Most of the above areas are south-facing slopes which because of the fact that the sun shines on them for most of the time have very little moisture retentivity. If suitable moisture conservation measures like contour trenching, terracing, closures to grazing are taken, the green cover on these slopes can be made to develop its own vegetable cover. Similarly in Kashmir Valley 1.30 lakh acres of land are *labroo* land, i.e., hill cultivated but hardly productive, with their top soil being washed away year after year. By terracing, gully plugging, suitable check dams, etc., soil erosion can be reduced and vegetable cover improved which will become a good source of energy. These degraded lands are ideally suited for energy plantation of trees.

Jammu Division

In spite of profuse precipitation the secret of stability of the Shiwaliks lay in its good forest cover but now this is a past story. The landslides which occur in Batote area, in Kashtwar-Batote road, in the entire Doda District are a direct result of gully erosion, tee erosion by nallah, and sheet erosion caused by unprofitable cultivation of roads, etc. The barren and waste areas of various districts of Jammu are as under:

Udhampur	—	46000	hectares
Jammu	—	77000	"
Doda	—	54000	"
Kathua	—	88000	"
Rajouri	—	45000	"
Poonch	—	12000	"

By suitable measures of terracing, gully plugging and contour terracing for moisture retention the above area can be developed to have a forest cover. Thus the future programme of energy production has to consist of (1) development of hydel energy, and (2) massive moisture conservation measures, soil conservation measures, afforestation and controlled grazing.

ANNEXURE I

Month-to-Month Rainfall in MM

Month	Jammu	Srinagar	Leh
January	65	70.9	0.9
February	53	73.4	7.9
March	51	94.2	7.1
April	20	90.2	5.8
May	24	59.7	5.6
June	54	94.3	0.6
July	313	56.1	11.9
August	334	62.2	15.9
September	49	35.3	6.9
October	14	21.2	2.5
November	7	13.7	10.2
December	31	37.3	4.8

ANNEXURE II

Rainfall coupled with the Snow Reserves

Month	Indus	Jehlum	Chinab
June	3283	16035	30816
July	5925	13847	42612
August	4369	10856	36320
September	1934	8248	17162
October	1409	4794	8425
November	906	3290	3265
December	640	2876	2585
January	525	2829	2197
February	560	4356	2099
March	573	8456	2597
April	742	14416	4601
May	878	17361	9221

ANNEXURE III

Total Hydel Energy of Hydel Projects being Investigated and Likely to be Taken up in the Next Decade

Sl. No.	Name of Project	Energy (in million units)	Cost (in Rs.lakhs)
1.	Upper Sindh Hydel Project	535.0	
2.	Nunwan-Batakot	152.8	3146.00
3.	Sheetkari Kulan	184.0	10018.00
4.	Boniyar	35.5	1254.00
5.	Athwathu	9.0	1711.00
6.	Brenwar	8.0	721.00
7.	Hirpora	40.0	1448.00
8.	Tangamarg	8.0	569.50
9.	Mutchil	0.16	—
10.	Asthan nallah	3.5	344.00
11.	Keran	3.6	173.00
12.	Wangat	6.0	—
13.	Ardoo Stage (I)	12.0	—
14.	Ardoo (II)	17.6	—
15.	Erin	15.8	—
16.	Pahalgam	15.8	700.00

NINE

ENERGY PROBLEMS AND PROJECTS IN J AND K STATE

Y.L. KHULAR

INTRODUCTION

Energy has played a vital and significant role in the lives of men even in primitive society while in the current technological era, its role has become crucial. In fact requirement of power and its availability has come to be recognized as the surest index of a country's overall economic scene because power is a basic input for industrial as well as agricultural achievements.

The Himalayas are the largest reservoir of snow and ice outside the polar regions and supports a multitude of glaciers. These glaciers act as natural sources for the supply of water to all the major river systems in north India. Although the economy of this region is dependent on the meltwater flow from these snow fields and glaciers, we have very little information about the causes of the seasonal streamflow variation and the total volumes of water being contributed from these sources. Apart from the influence of the glacier melt flow on our agricultural economy and environment, the glacier contributions in the upper and lower reaches can be used for the generation of cheap hydro-electric power (potential estimated to be around 100,000 MW).

It is estimated that the glaciated area in the Himalayas is over 33,200 sq. km. whereas seasonal snow covers over 10 times this area during winters. Glaciers occupy about 17 per cent of the total mountain area in the Himalayas. A rough estimate shows that there are over 15,000 glaciers in these high mountain regions. The Himalayas'

East-West orientation and high latitude and altitude make them a unique system in the world. The criss-cross chain of mountain ranges, i.e. Zanskar Range, Ladakh Range, Pir Panjal Range, Dhaula Dhar, etc., in the Punjab Himalayas with different orientations and elevations provides a wide range of climatic systems. The Himalayan glaciers could be broadly described to belong to:

(i) Predominantly Monsoonal (Eastern Region)
(ii) Dominant winter precipitations (Western Region)
(iii) The Central Region is affected by both the Monsoon and Westerly disturbances.

There is need to collect data for detailed studies in these regions to study the glacier-atmosphere, interrelationships and the climatic change. The precipitation in the Himalayas changes significantly with altitude and location. The maximum rainfall occurs near the foothills of the Himalayas. As one proceeds northwards, the rainfall decreases. Rainfall, however, again increases in the Himalayas at the elevation of 2,000 to 2,400 mm. Himalayan glaciers have a large variation in their size which ranges from small glaciers at high altitudes to enormous ice flows coming down from wide basins.

Jammu and Kashmir State in particular has plenty of Hydel capacity from the various river systems and small streams which are snow-fed in terms of glaciers and seasonal snow cover melt as well as seasonal rains. The river system, viz. Indus, Zanskar, Nubra, Shyok, Suru, Dras, Sindh, Liddar, Krishan Ganga, Jhelum, Pohru, Chenab, Marau, Tawi, Naoshera Tawi, Punch, etc., flowing in J&K, if properly and scientifically exploited, can provide enough, rather surplus, power to the country and in particular for Jammu and Kashmir. So far exploitation of the huge hydel resource has been minimal/marginal. In addition to a number of medium and large hydel projects, the State could generate plenty of power from micro/hydel schemes from run-off of the small streams being fed by glaciers, seasonal snow and rains.

Hydel Resources in J & K

Prior to 1947, there was hardly any hydel generation in the State except a minor hydel project of Mohura Hydel Station. Thereafter, serious and sincere efforts have been made to exploit water resources by constructing Ganderbal Hydel Project, Chineni Hydel Project, Upper Sindh Hydel Project and Lower Jhelum Hydel Project whereas Salal Hydel on the Chenab, Dulhasti on the Chenab, Lower Sindh in Kashmir Valley are under various stages of construction. In addition it is also seen that a few other micro-hydel schemes have also been executed in the State.

In the past two decades, surveys for feasibility studies of many major and medium hydel projects have been carried out for the Power Development Department of the State, by the survey of India viz. :

(i) Krishan Ganga Cross Drainage Hydel Project
(ii) Panchtarni Storage & Hydel Project
(iii) Liddar Hydel Project
(iv) Uri Hydel Project
(v) Dras Hydel Project
(vi) Tangtse Hydel Project
(vii) Hanzal Hydel Project Marau
(viii) Pakhal Hydel Project Marau
(ix) Bhandarkot Hydel Project Marau
(x) Anas Hydel Project
(xi) Ratle Hydel Project
(xii) Sewa Hydel Project
(xiii) Gaik Hydel Project
(xiv) Ikhale Hydel Project
(xv) Ujh Hydel Project
(xvi) Gangabal Hydel Project
(xvii) Nagar Dam Hydel Project
(xviii) Gawar Dam Hydel Project
(xix) Bursar Dam Hydel Project
(xx) Banzwar Hydel Project
(xxi) Baglihar Hydel Project & few others.

Thereafter I am sure that the Government of Jammu and Kashmir in these Five Year Plans must be looking forward and catering for more and more hydel generation.

Feasibility of any Micro, Medium and Major Hydel Schemes can be studied by utilizing 1:50000 or 1:25000 maps already available which depict the extent of glaciers and help the planner to estimate accurately the seasonal snow cover. Almost all streams are fed to the extent of 25 per cent of total run-off/snowmelt in March-April-May period and 45 per cent of the total run-off in June-July-August-September period each year, and the remaining 30 per cent during October to February indicating that glacier-fed streams are not only perennial but can also be modulated with storage/reservoir to generate 50 per cent of the installed capacity even in the lean period.

Alternate Energy Sources Excluding Solar Power

Jammu and Kashmir can ill-afford to have thermal generation on account of inadequate availability of coal. The transportation of coal all the way from the coal-mining areas is not at all profitable. The cost of thermal power generation in the plains of Punjab/Haryana is double that of hydel generation as such if J&K has to undertake thermal generation even for balancing in the lean period, the cost of generation may be two and half times or more of the cost of the hydel generation which people in general and industry in particular may not be able to bear.

The next alternative can be nuclear power. This can be thought over and feasibility for cost-benefit ratio worked out primarily for balancing the power needs during the lean period of winter, when hydel capacity is reduced. The location and capacity of nuclear power plants have to be in keeping with the overall power policy of the Government and the security hazards on account of the vicinity of the border with Pakistan and vulnerability to ground and air attacks.

River System for Hydel Generation in J&K

The Hydel Projects, Thermal Generation Projects and Nuclear Power Projects need funds for execution, fitting in with the National Power Policy and suitable to the Northern Grid.

The rivers in J&K are fed from big and small glaciers of the Karakoram mountains and of Ladakh Ranges, Zanskar Range, the great Himalayas and Pir Panjal Ranges.

Rivers Nubra and Shyok drain out water from Karakoram mountains and Ladakh Ranges whereas the Indus and Zanskar drain water from Ladakh and Zanskar Ranges.

Suru river drains out water of the northern slopes of Nunkun Peak and Glaciers in that Range, Jhelum drains out water from the northern slopes of the Pir Panjal Range and watershed range dividing the watershed of Jhelum with Marau the tributary of river Chenab.

River Sindh drains out water from the snowbound areas of Panchtarni and adjoining glaciers and snowmelt of SW and western slopes of the watershed from the Zogila Pass towards Amarnath caves and along the Ridge.

River Liddar east and west, drains out water from the snowfed area of Sheshnag glacier and adjoining areas and Kolhai Glaciers. River Krishan Ganga drains out glacier and snow-melt of the Northern slope of watershed known by Sogpu Range including Talel valley.

It will thus be seen that there are many rivers which are fed with plenty of glaciermelt and seasonal snowmelt. Few of the streams which drain out water from the Himalayan Range also drain out seasonal rain contribution. Streams/rivers falling in the Kargil and Ladakh area generally do not have any contribution from seasonal rains as there are few rains in the area north of the great Himalayan Range.

Hydel Resources of Punjab Himalayas

The Punjab Himalayas starting from the sharp turning of river Indus near the Nanga Parbat up to the crossing of the Satluj near Shipki La is 560 km in length and 100 to 150 km in depth including the Greater Himalayas, Lesser Himalayas and Lower Himalayas. The Punjab Himalayas are drained in the northern face by the Indus and in the south face by the Krishan Ganga, Jhelum, Chenab, Ravi, Beas and partly by the Satluj. In these Punjab Himalayas all the area, about 1,700 m/2,000 m above Mean Sea Level, has plenty of seasonal snowfall depending on altitude/latitude and location north or south of the Great Himalayan Range and thus contribute heavily in the run-off of the rivers during April-September every year.

Whole of the Himalayas are estimated to yield snow-melt contribution of 700^3 km whereas the Punjab Himalayas are likely to contribute about 200/250 km^3 – at a rough estimate. Further, against the overall average of 17 per cent Himalayas, i.e. about 33,200 sq. km. being glaciated, the Punjab Himalayas are estimated to have a higher percentage of glaciated covers on account of higher altitude and high latitude besides being served by N/W winds/storms during the winter months bringing about heavy precipitation.

For the sake of discussion, it was seen that approximately 705 sq. km., 206 sq. km., 170 sq. km., 3,000 sq. km., 2,000 sq. km., 8,800 sq. km. of catchment areas of rivers Beas, Ravi, Jhelum, Chenab, Satluj and Indus respectively are covered by permanent snow/glaciers as could be roughly worked out from the map coverage available.

Glacier and Catchment Sheds for Hydel Energy

We have, in the past decade, in association with the Geological Survey of India, Meteorological Department, WADIA Institute of Himalayan Geology, Indian Space Research Organization, Department of Science and Technology, etc., carried out studies of various glaciers, for preparing an inventory of glaciers, size, shape and lay of glaciers seasonal snowcover extent, precipitation depth at different altitudes, latitudes and Ranges. Mass balance studies, hydro meteorological parameters, i.e., net radiation, humidity temperature, wind direction/speed, moisture precipitation and snow water equivalent, for scientific analysis and forecasting yield from various glaciers by field observation, map studies, aerial photographs and satellite imagery interpretation. These studies were undertaken to do in-depth studies of glaciers and fix up various parameters which contribute to the behaviour of glaciers and snowmelt. So far we have either been depending on continental parameters or empirical methods to forecast snowmelt contribution in various streams/hydel project reservoirs.

We have, in recent past, carried out catchment studies of rivers Beas and Satluj. The catchment of the Beas above the Pandoh Dam was taken up. We find that above 700 sq. km. of area has permanent snow and 3,700 sq. km. area is above 2,000 m altitude and 930 sq. km. of area is below 2,000 m above Mean Sea Level. Again from Beas river between the Pandoh and Pong Dams 1,130 sq. km. of area is above 2,000 altitude and 5,900 sq. km. is below 2,000 m above Mean Sea Level which gives a fairly good idea of permanent snow-melt and seasonal snow-melt contribution in the Beas for forecasting the pattern of flow in Pandoh and Pong Dam Hydel Projects.

Similar studies were done for the Satluj catchment and found that 2,000 sq. km of permanent glacier exists in the Indian and Tibetan areas and it was further found that 15,000 sq. km. of area is above 2,000 m altitude and 4,660 sq. km. of area falls below 2,000 m above Mean Sea Level in India.

We had not carried out any such studies for the J&K rivers as we were never requested for such information. We have, however, rough figures of glaciated areas and catchment areas of rivers as below:

		Glaciated Area/Catchment
Beas	—	705/14500 sq. km.
Ravi	—	206/8000 sq. km.
Jhelum	—	170/33600 sq. km.
Chenab	—	3000/27200 sq. km.
Satluj	—	2000/48000 sq. km.
Indus	—	8800/268800 sq. km.

Areas about 1700-2000 m above Mean Sea Level which is almost 10 times of the Glaciated area in these catchments can be worked which gives a fairly good idea of seasonal snowcover.

However, a detailed study of the catchment of river Krishan Ganga falling in the Indian side of L.O.C. Sindh, Liddar and Marau provide the following useful data:

	Glacier Area	Catchment	Seasonal Snow Cover
Krishan Ganga	30 sq. km.	1570 sq. km.	1500 sq. km.
Sindh	32 sq. km.	1430 sq. km.	1400 sq. km.
Liddar	66 sq. km.	1100 sq. km.	1000 sq. km.
Marau	841 sq. km.	4580 sq. km.	4000 sq. km.

Conclusion

There is a chronic power shortage in the State of Jammu and Kashmir which can be overcome only by augmenting hydel generation, as other alternatives are rather uneconomical except for the limited purposes of balancing power requirements during the lean period of hydel generation in the winter months. The availability of a huge hydel potential in the Punjab Himalayas and in particular in the J&K should, therefore, be fully tapped. For this detailed scientific studies of utilization and exploitation of these resources should be carried out by the appropriate Departments of the Government of Jammu and Kashmir in close liaison with the Survey of India with a view to planning and executing Hydel Projects in the State on a long-term and short-term basis for the next three decades. Simultaneously, the power transmission line capacity should be so planned as to take future Hydel

Generation en route, tagged along from various Major, Minor and Micro Hydel Projects in the interest of overall economy and faster cost-benefit returns at minimal investments.

REFERENCES

1. Bahadur, J., *Himalayan Water from Snow-Ice & Glacier.*
2. Brig, Y.L. Khular, *The Great Himalayas background paper.*
3. Edwin T., Atkinson, *The Himalayan Gazetteers.*
4. Kyuya Fikada, *The Great Himalayas.*
5. Lall J.S. and Bahadur J., *The Himalayas aspect of change Himalayan Environment and Devlopment of its Perennial River Systems.*

TEN

PLANNING FOR FOREST ENERGY SUPPLY IN JAMMU AND KASHMIR

MOHAMMAD SULTAN BHAT & SHAH NAWAZ ALAM

Energy is one of the vital preconditions of life. Growth cannot be comprehended without understanding the laws of energy. The rise and decay of human civilizations have closely been linked with the rise and fall in per capita consumption of energy. Controlled use of fire mothered industrial civilization and man started drawing energy from fossil fuels. Thus started the process of man eating into nature's capital, a process of self-annihilation. Today this process has assumed staggering dimensions. Aurelio Peccei (founder President of the Club of Rome) in his address to the Club while referring to the environmental crisis says:

> We should in this regard know that the state of our planet is none too good. It offers a dismaying picture of depletion and degradation: Wilderness, the treasure chest of nature, disappearing; deserts advancing; tropical forests rapidly destroyed; coastal zones and estuaries ruined; large numbers of animal and plant species condemned to extinction; waters, soils and the very air we breathe contaminated; natural cycles, climate and other ozone layer tempered with, often irreversibly; human ecology directly effected by over-harvesting, over-grazing and over-fishing which submits the strategic biological systems of croplands, pasture lands and fisheries to stress precisely when they are called upon to satisfy exponentially growing human demands. And so on and so forth.
>
> (*Seminar*, June 1985)

The predicament of modern man is how to reconcile, in our single world, the energy requirements of those "who have plenty and still claim more with the needs of those who desperately want just to stay alive and not fall into even more abject poverty". The world population, which is now at record high of 4.7 billion, is expected to exceed 6 billion in the year 2000, thus adding more people in barely 15 years than those living on the earth one century ago. We thus face a real question as to whether there are enough energy resources to maintain the living standards in the developed countries and improve the living standards of the people, even modestly, in the less developed nations. In search for solutions to energy challenges it has been emphatically argued that scientific exploitation of renewable energy sources provides an answer to an ever-increasing demand for energy.

Wood can decidely make a distinct contribution in this direction. "The world's forests, covering some 40 million km., contain about 340 million cu.m. of timber. The value of the world's forest products is today 115 billion U.S. dollars per year. At present only 10 per cent of the world's surface is used for food production. With the aid of trees three quarters of the globe could supply food, fibre, fuel and shelter. We can use more wood without cutting additional trees. Technology is showing us how to do it. Recent studies for the U.S. Energy Research and Development Administration point to future possibilities of changing the physical nature of wood on a large scale in order to produce wonder products like supertree, redwood, naturals gas, L-Dopa, Phenol, Benzene, Ammonia, Formaldehyde, etc., from it. It has been demonstrated that one forest acre, left to nature, would grow a stand of mature trees in 250 years. The same acre, with the help of scientific farming, could grow a stand ready for cutting after every 50 years – thus providing five harvests instead of one in the same time, with six times the volume of wood of unmanaged forests." Moreover, forest farming generates a large array of externalities. Watershed forests, for example, are not only a source of timber but preserve soil cover on site, protect soil erosion, smooth the supply of water for agriculture from irregular rainfall and prevent water reservoirs and irrigation systems from sedimentation. Similarly tropical rain forests, in addition to the above externalities, provide a habitat of diverse genetic pool, myriad of living species, which are extremely necessary for the balanced functioning of the eco-system.

India's opportunities of tapping its forest resources for facing the twin challenges of eco-development and energy are immense. In the face of uncertain alternative energy supplies for forest fuel or even uneconomical substitutions our approach to forest management needs to be reformulated. Unfortunately we have exhausted our forest wealth more extensively without learning how to rebuild it better and faster. According to a report of the National Remote Sensing Agency – released in November 1984 – there has been 16.25 per cent net reduction of forest cover between 1975 and 1982. There are only 33 million hectares with 30 per cent density against the time-honoured figure of 75 million hectares of forest cover in the country. This is a disturbing situation as ecologists maintain that when a natural forest is thinned to below 60 per cent of its tree population it ceases to perform its ecological functions. At present, there are 175 million hectares of degraded land which speaks for the vast scope of increasing tree and grass cover.

Forests can sustain the required supply of fuel wood only if we bring their management under the purview of a well defined long-term energy policy. This is necessary if needless liquidation of much of the existing forest capital is to be avoided. The greatest cause of deforestation is not the market for wood products but rather the need for wood fuel and agricultural land. It has been estimated that 50 per cent of the trees cut in the world are for fuel use. Some 12 to 24 million acres of forest area are depleted annually for agriculture alone. Thus in a rational land-use pattern and long-term energy policy forests have a definite role and importance. It is against this backdrop that an attempt will be made in the following:

(i) Delineate the role of forests in the energy planning of Jammu and Kashmir State,
(ii) Review the existing forest management of the State,
(iii) Present a profile of planning for the forest energy supply for the State in accordance with the long-term planning objectives of the State.

Forest Resources of the States

Forestry constitutes one of the biggest and the most productive industries of the State economy. It acts as a necessary adjunct to a number of forest-based industries. Horticulture and tourist industries also depend upon forest industry in a variety of ways.

The area[1] under forests constitutes 15.19 per cent of the total geographical area of the State. Ladakh District, which covers nearly 70 per cent of the State's territory, is a high plateau devoid of forests. Hence if Ladakh is excluded from this calculation the rest of the State has nearly 50 per cent of its area under forests. The figure, though not up to the 66 per cent prescribed for the Himalayas and other hilly areas, is quite high as compared to many other States. The State's per capita forest area of 0.35 hectares is quite well above the national average of 0.12 hectares.

Planwise analysis of the forest area data shows that during the First, Third and Fourth Five Year Plan periods area has remained constant. During the Second, Sixth and Fifth Five Year Plan periods it recorded a growth rate of -9.15, -3.10 and 1.44 per cent respectively. The decadal trend growth rates (i.e. 1954-55 to 1963-64, 1964-65 to 1973-74 and 1974-75 to 1983-84) work out to be -3.53, -2.80 and 0.85 per cent respectively. The overall picture from 1951-52 to 1983-84 shows a net reduction in the forest area to the tune of -0.76 per cent per annum (Table 10.1).

TABLE 10.1: **Plan-wise Decadal Trend Growth Rates of Area under Forests in Jammu and Kashmir State**

Plan/Decade	Linear Function		Log-Linear Function		Annual Growth Rate
	Regression Coefficient (sq. km.)	t value	Regression Coefficient (Sq. Km.)	t value	
1	2	3	4	5	6
I Plan++	—	—	—	—	—
II Plan	–2353.611**	2.4380	–0.0915*	6.2954	–9.15
III Plan++	—	—	—	—	—
IV Plan++	—	—	—	—	—
V Plan	+313.418	2.2359	+0.0144*	7.1744	+1.44
VI Plan	–685.056**	4.0000	–0.0310**	4.3335	–3.10
1954-55 to 1963-64	–1394**	4.3956	–0.0353*	5.6186	–3.53
1964-65 to 1973-74	–6.643*	5.5078	–0.0280*	27.6100	–2.80
1974-75 to 1983-84	+183.050***	1.9611	+0.0085*	28.8488	+0.85
Overall (1951-52 to 1983-84)	–195.282*	4.1483	–0.0076*	291.89	–0.76

* Significant at 1 per cent level
** Significant at 5 per cent level
*** Significant at 10 per cent level
++ Constant

The existing production from forests is mainly in the form of timber, fuelwood, charcoal, minor forest products and different varieties of fodder. The total stock of commercial forests in the State is 981,179 thousand cu.m. The coverage extends over 502 thousand hectares which works out to be 40.50 per cent of the total forest area. The growing stock per hectare in the State is 175 cu.m. as compared to 32 cu.m. at the national level and 110 cu.m. for the world as a whole. The average yield per hectare of the commercial forest area in the State is 1.78 cu.m. Forests of the State are predominantly coniferous-Deodar, Kail, Fir, Chir are economically more valuable tree species. About 2,051,000 hectares are covered by permanent forests and only 34,000 hectares are under unclassed forests. Total land under pastures in the State is 4,193.53 sq.km. Although a small percentage of the State's livestock population is able to graze in the forests the damage caused by migratory grazers is serious. The climate of Kashmir, with its great altitudinal variations ranging from 150 m. to 4,500 m., is conducive to the growth of a wide range of minor forest products particularly medicinal plants. The Kashmir forests are the richest as far as such products are concerned.[2] Commercial value of these products came to be recognized with the increasing popularity of the *Yunani* and *Ayurvedic* systems of treatment. Output of timber and firewood, the major components of primary forest products, has recorded an annual compound growth rate of 2.89 per cent and -0.37 per cent respectively from 1951-52 to 1983-84 (Tables 10.2 and 10.3). During the decades 1954-55 to 1963-64, 1964-65 to 1973-74 and 1974-75 to 1983-84 output of firewood has recorded a growth rate of -2.37, -7.60 and 9.79 per cent respectively. The growth rate of timber and firewood put together works out to 1.60 per cent per annum from 1951-52 to 1983-84 (Table 10.4). Minor forest products recorded a growth rate of 19.55 per cent per annum over the period of past 33 years from 1951-52 to 1983-84 (Table 10.5).

Due to the non-availability of adequate, cheap and reliable sources of fossil fuels, firewood, dung, sawdust, dry leaves, charcoal, soft coke, kerosene and cooking gas are the main sources of energy for domestic consumption. The major sources of fuelwood are:

1. Government
2. Owned sources
3. Purchased
4. Other sources:

TABLE 10.2: **Plan-wise/Decadal Trend Growth Rates of Outturn of Timber in Jammu and Kashmir State**

Plan/Decade	Linear Function		Log-Linear Function		Annual Growth Rate
	Regression Coefficient lakh/cft	t value	Regression Coefficient	t value	
1	2	3	4	5	6
I Plan	15.918**	2.4831	+0.2294*	7.4964	+22.94
II Plan	0.215	0.0505	+0.0018*	7.0352	+0.18
III Plan	–7.910***	1.6722	–0.0607*	6.5553	–6.07
IV Plan	–5.970	1.1860	–0.0453*	6.6581	–4.53
V Plan	+11.292**	3.0563	+0.0759*	7.5607	+7.59
VI Plan	–2.761*	8.8267	–0.0130**	4.4143	–1.30
1954-55 to 1963-64	+4.4389**	2.0956	0.0411*	160.1720	+4.11
1964-65 to 1973-74	–0.4912	0.2351	–0.0032*	27.4614	–.032
1974-75 to 1983-84	+8.4388*	6.2886	+0.0551*	29.9748	+5.51
Overall (1951-52 to 1983-84)	+3.2774*	5.7731	+0.0289*	28.0371	+2.89

* Significant at 1 per cent level
** Significant at 5 per cent level
*** Significant at 10 per cent level

TABLE 10.3: **Plan-wise/Decadal Trend Growth Rates of Outturn of Firewood in Jammu and Kashmir State**

Plan/Decade	Linear Function		Log-Linear Function		Annual Growth Rate
	Regression Coefficient (lakh/cft)	t value	Regression Coefficient	t value	
1	2	3	4	5	6
I Plan	–1.682	0.3101	–0.044*	6.1047	–4.40
II Plan	–0.184	0.0857	–0.0021*	5.0992	–0.21
III Plan	–2.422***	2.1180	–0.0702*	5.1145	–7.02
IV Plan	–2.516**	3.1371	–0.0868*	5.4654	–8.68
V Plan	+3.350**	4.0449	+0.0956*	7.7161	+9.56
VI Plan	–12.942**	2.3565	–0.1805**	3.6584	–18.05
1954-55 to 1963-64	–1.2370	1.1296	–0.0237*	24.9065	–2.37
1964-65 to 1973-74	–0.7457**	2.0402	–0.0760**	3.5703	–7.60
1974-75 to 1983-84	+4.0330**	2.7657	+0.0979*	28.4454	+9.79
Overall (1951-52 to 1983-84)	–0.1072	0.3642	–0.0037*	18.2476	–0.037

* Significant at 1 per cent level
** Significant at 5 per cent level
*** Significant at 10 per cent level

TABLE 10.4: **Plan-wise/Decadal Trend Growth Rates of Total Major Forest Products in Jammu and Kashmir**

Plan/Decade	Linear Function		Log-Linear Function		Annual Growth Rate
	Regression Coefficient (lakh/cft)	t value	Regression Coefficient	t value	
1	2	3	4	5	6
I Plan	+14.236*	5.5626	0.1184*	7.8872	+11.84
II Plan	+0.038	0.0097	0.0008*	7.0528	+0.08
III Plan	−10.341**	2.6180	−0.061*	6.5992	−6.10
IV Plan	−8.488***	1.6845	−0.052*	6.6324	−5.29
V Plan	+14.631**	3.7693	+0.0796*	7.6034	+7.96
VI Plan	−7.853**	2.7292	−0.028**	4.3435	−2.80
1954-55 to 1963-64	+3.2021	1.1345	+0.0287*	5.0420	+2.87
1964-65 to 1973-74	−1.2287	0.5875	−0.0005*	27.8436	−0.05
1974-75 to 1983-84	+12.393*	5.0070	+0.0590*	7.4024	+5.90
Overall (1951-52 to 1983-84)	+3.1702*	4.5818	+0.0160*	41.4699	+1.60

* Significant at 1 per cent level
** Significant at 5 per cent level
*** Significant at 10 per cent level

TABLE 10.5: **Plan-wise/Decadal Trend Growth Rates of Outturn of Minor Forest Products in Jammu and Kashmir**

Plan/Decade	Linear Function		Log-Linear Function		Annual Growth Rate
	Regression Coefficient (lakh Rs.)	t value	Regression Coefficient	t value	
1	2	3	4	5	6
I Plan	−0.649	1.1320	−0.0525*	6.5030	−5.25
II Plan	−2.201***	2.3073	−0.0887*	6.3369	−8.87
III Plan	+0.932	1.3392	+0.0380*	7.2847	+3.80
IV Plan	+65.498	1.1683	+0.8242*	12.4694	+82.42
V Plan	+335.56*	9.7255	+0.4064*	9.7075	+40.64
VI Plan	−136.02**	2.4732	−0.0980**	4.0106	−9.80
1954-55 to 1963-64	+1.058**	1.9409	+0.0782*	3.1600	+7.82
1964-65 to 1973-74	+20.967*	3.0113	+0.2626*	22.3800	+26.26
1974-75 to 1983-84	+75.300***	1.8181	+0.0879*	24.1225	+8.79
Overall (1951-52 to 1983-84)	51.740*	7.4332	+0.1955	0.7911	+19.55

* Significant at 1 per cent level
** Significant at 5 per cent level
*** Significant at 10 per cent level

Tree groves in :

(a) village common land
(b) orchards
(c) agricultural land
(d) grazing land
(e) revenue land, and
(f) other waste lands.

According to a Pilot Wood Balance Study[3] firewood is the most important source of energy in the rural areas of the State though its distribution according to income, area and seasons is skewed within rural areas. Consumption of kerosene and gas is, by and large, confined to urban areas. Per-capita-per-annum consumption of firewood, twigs/branches, charcoal, sawdust, dung, dry leaves and kerosene for the State as a whole has been worked out as 3.8, 1.6, 0.2, 0.2, 1.0, 0.2 (quintals) and 8.7 (litres) respectively. For the rural and urban areas of the State the corresponding figures are 4.4 quintals and 2.5 quintals, 2.1 and 0.4, 0.2 and 0.1, 0.1 and 0.4, 1.3 and 0.3, 0.3 and 0.9, 4.0 and 19.8 litres respectively. The species-wise consumption of firewood indicates that 41 per cent of the total firewood originates from twigs/branches, 37.50 per cent from coniferous wood, 10 per cent from willow and rest by *Kiker, Hatab, Walnuts, Poplar, Zangloo, Fruit tree, Banjmuaroo, Tali Eucalyptus,* etc. (74.2 per cent of the total consumption of domestic fuel comprise firewood). Firewood and dungcake together constitute 94.7 per cent of the total consumption of energy of the rural households. In the urban areas firewood, kerosene, sawdust and cooking gas account for 90 per cent of the domestic consumption of energy. Consumption of firewood is negatively related to income as per the results of the above referred survey. Wood is also consumed by bakers, barbers, hoteliers, milk vendors and is required for mosques, temples, social ceremonies and cremation. Demand on this account is likely to increase as a result of increase in population and expansion of the tertiary sector of the State. *Though it will be erroneous to derive any reliable conclusion from the said study – precisely because its limited scope and methodological issues – still the results of the study are indicative of the fact that wood shall continue to dominate the consumption pattern of energy in the State for decades to come.*

According to an unpublished report of the Jammu and Kashmir

Forest Department,[4] "approximately 50,000 tonnes of fuelwood is produced annually from the State forests. Commercial energy sources (coal, electricity, kerosene, etc.) meet about 3 per cent of the energy requirements. The remaining 97 per cent is met from fuelwood and dung-cakes. Thus out of the total domestic energy consumption of 6.2 m. tonnes in the State, 0.2 m. tonnes is met from commercial energy sources and for the remaining 6 million tons fuelwood and dung-cake are used, resulting in heavy pressure on natural forests".

Revenue from Forests

The contribution of forests to the State Domestic Product (SDP) and revenue has markedly increased over the years. In 1947-48 net surplus from forests was Rs. 76.68 lakhs and in 1970-71 contribution from this sector to SDP was Rs. 12 crores. By the end of the year 1979-80 it exceeded Rs. 50 crores. Similarly, in 1947-48 forest revenue amounted to Rs. 103.10 lakhs, in 1965-66 it reached Rs. 3 crores. In 1983-84 revenue from forests amounted to Rs. 34 crores.

From 1951-52 to 1983-84 the annual trend growth rate of forest revenue works out to be 12.53 per cent. For the First, Second, Third, Fourth, Fifth and Sixth Plan periods the growth rates work out to be 11.14, 20.71, -2.47, 7.33, 24.79 and -0.77 respectively (Table 10.6). Decadal growth rates of forest revenue have been estimated as 20.06, 9.14, and 19.94 for the decades 1954-55 to 1963-64, 1964-65 to 1973-74 and 1974-75 to 1983-84 respectively. Net surplus from forests has recorded a 9.72 per cent annual growth rate from 1950-51 to 1981-82 (Table 10.7). Corresponding figures for the First, Second, Third, Fourth and Fifth Five Year Plan periods are 12.53 22.74, -9.68, 9.47 and 11.4 respectively. From 1952-53 to 1961-62, 1962-63 to 1971-72 and from 1972-73 to 1981-82 the growth rates of net surplus from forests work out to be 28.38, 1.43, and 13.94 per cent annually respectively. Forest revenue as a percentage of State revenue has declined from 12.4 per cent to 3 per cent between 1949-50 and 1980-81 (*Digest of Statistics*, 1982).

Expenditure on Forest Development

The expenditure on development of forests has been mounting with every Five Year Plan from Rs.160.2 lakhs in the First Five Year Plan to Rs. 1211.98 lakhs in the year 1981-82. The plan-wise growth rates

TABLE 10.6: **Plan-wise/Decadal Trend Growth Rates of Revenue Receipts from Forests in Jammu and Kashmir State**

Plan/Decade	Linear Function		Log-Linear Function		Annual Growth Rate
	Regression Coefficient (lakh Rs.)	t value	Regression Coefficient	t value	
1	2	3	4	5	6
I Plan	+8.488*	6.1440	+0.1114*	7.8438	+11.14
II Plan	+47.120*	6.7411	+0.2071*	8.4290	+20.71
III Plan	–8.139	0.9155	–0.0247*	92.0098	–2.47
IV Plan	+42.565**	4.0345	+0.0733*	7.5692	+7.33
V Plan	+293.035*	5.0334	+0.3479**	8.6055	+24.79
VI Plan	–26.569	0.5810	– 0.0077**	4.4360	–0.77
1954-55 to 1963-64	+37.1230*	12.3496	. 0.2006*	32.3600	+20.06
1964-65 to 1973-74	+38.7300*	7.9330	+0.0914*	30.7447	+9.14
1974-75 to 1983-84	+352.35*	9.8510	+0.1994*	32.2400	+19.94
Overall (1951-52 to 1983-84)	+95.54*	8.5056	+0.1253*	212.33	+12.52

* Significant at 1 per cent level
** Significant at 5 per cent level
*** Significant at 10 per cent level

TABLE 10.7: **Plan-wise/Decadal Trend Growth Rates of Net Surplus Derived from Forests in Jammu and Kashmir State**

Plan/Decade	Linear Function		Log-Linear Function		Annual Growth Rate
	Regression Coefficient (lakh Rs.)	t value	Regression Coefficient	t value	
1	2	3	4	5	6
I Plan	+5.847**	2.7195	+0.1253*	6.0996	+12.53
II Plan	+39.322*	4.6636	+0.2274*	6.6690	+22.74
III Plan	–23.058***	1.8285	0.0968*	4.8511	–9.68
IV Plan	+31.761**	4.1717	+0.0947*	5.9787	+9.47
V Plan	+46.778*	4.8601	+0.1140*	6.0776	+11.40
1952-53 to 1961-62	+30.917*	10.1794	+0.2838*	31.1906	+28.38
1962-63 to 1971-72	+3.593	0.6073	+0.0143*	4.1201	+1.43
1972-73 to 1981-82	+182.440*	4.5078	+0.1394*	12.9190	+13.94
Overall (1950-51 to 1981-82)	+36.344*	5.6298	+0.0972*	53.2638	+9.72

* Significant at 1 per cent level
** Significant at 5 per cent level
*** Significant at 10 per cent level

of forest expenditure for the First, Second, Third, Fourth and Fifth Five Year Plan periods work out to be 8.74, 12.40, 15.00, 3.41 and 43.05 respectively (Table 10.8). During the Sixth Plan period this figure will certainly be on the higher side due to the Rs.24 crores World Bank financed Social Forestry Project. The Decadal Growth Rates (from 1952-53 to 1961-62, 1962-63 to 1971-72 and 1972-73 to 1981-82) have been estimated as 12.33, 12.92 and 23.14 per cent respectively while the overall rate of growth comes out to be 13.29 per cent per annum from 1950-51 to 1981-82. Expenditure on "forest development proper" as a percentage of forest expenditure has increased from 0.075 in 1949-50 to 20.44 in 1975-76 (the year for which comparable figures are available). Though *normal expenditure* as a percentage of total expenditure on forests has declined from 99.92 in 1949-50 to 79.55 in 1975-76 still this percentage is on a much higher side compared to the expenditure on *forest development proper*.

TABLE 10.8: **Plan-wise/Decadal Trend Growth Rates of Expenditure on Forests in Jammu and Kashmir State**

Plan/Decade	Linear Function		Log-Linear Function		Annual Growth Rate
	Regression Coefficient (lakh Rs.)	t value	Regression Coefficient	t value	
1	2	3	4	5	6
I Plan	2.641*	4.9830	0.874*	5.9487	+8.74
II Plan	7.802*	5.3888	0.1240*	5.8852	+12.40
III Plan	14.900*	10.9478	0.1500*	6.2979	+15.00
IV Plan	8.311	0.4698	0.0341*	90.7183	+3.41
V Plan	233.529***	3.0438	0.4305*	6.3607	+43.05
1952-53 to 1961-62	5.805*	11.1398	0.1233*	12.7794	+12.33
1962-63 to 1971-72	20.0747*	15.1678	0.1292*	9.9232	+12.92
1972-73 to 1981-82	124.818*	5.5823	0.2314*	4.2080	+23.14
Overall (1950-51 to 1981-82)	30.747*	7.3228	0.1329*	254.01	+13.29

* Significant at 1 per cent level
** Significant at 5 per cent level
*** Significant at 10 per cent level

In the First Five Year Plan, soil conservation schemes were accorded top priority and more than 60 per cent funds provided were spent on soil conservation schemes. Forestry schemes were nominal. The Second and Third Five Year Plans involved, by and large, continuation and enlargement of work taken up during the First Plan. The main achievements of the Fourth Plan were bringing 3,245 and 1,391 hectares under fuelwood and plantation respectively. The Fifth Plan marked a definite shift both in the pattern of expenditure and focus of the policies pursued for protection and preservation of forests.

Management of Forests

The history of the Forest Department as an organized body commenced from the year 1891 A.D. when McDonnel joined as a Lent officer from the Government of India. Before 1891 A.D. forests were exploited in a haphazard manner. The existing management of forests owes much to the pioneering work of Diwan Amar Nath Kaul, J.C. McDonnel, Syed Mehdi Hassan, Lovegrove, B.O. Conventry, Dr. Simson, H.L. Wright, C. S. Martin, Sir Peter Clutterbuck and Harman Singh Pathania.

Till 1947 the most important systems of exploitation in the State were *Selection felling*, *Coppice Felling* and *Unregulated Felling*. The greater part of *Deodar, Kail and Chir* was extracted according to Selection Felling. Hardwood and firewood came from Coppice Felling while fir was acquired by Unregulated Felling. No less than 60 per cent of the total out-turn used to come from Unregulated Felling. Though forests in Jammu and Kashmir have been brought under "scientific management" since 1931 yet the process of organized forest development on an extensive scale started only from 1951-52. At present forests in the State are managed according to the following systems:

1. Uniform System
2. Selection System
3. Protection-cum-Improvement System
4. Unregular System
5. Mixed Coniferous Selection System, and
6. Fir Selection System.

Each system is further sub-divided into commercial and uncommercial categories. Area under each system as a percentage of the total area (area under all categories) is 11.14, 17.22, 28.34, 25.66, 9.11 and 8.53 respectively. Area under commercial and uncommercial sub-categories as a percentage of area under each system is 83.92 and

16.08, 63.61 and 36.39, 31.08 and 68.92, 15.33 and 84.63, 71.69 and 28.30, 51.42 and 48.58 respectively (Table 10.9). Out of the total area system managementwise only 46.64 per cent is under commercial forests and the rest is under uncommercial forests. Methods of harvesting continue to be primitive and a lot of valuable lumber is wasted. Methods of sawing, transportation and protection have undergone encouraging improvement:

TABLE 10.9: **Percentage Area by Systems of Management in Jammu and Kashmir State**

Uniform			Selection			Protection-cum-Improvement		
1	2	3	4	5	6	7	8	9
Commer-cial	Un-com-mercial	S. Total	Commer-cial	Un-com-mercial	S. Total	Commer-cial	Un-com-mercial	S. Total
83.92	16.08	11.14	63.61	36.39	17.22	31.08	68.92	28.34

U-regular			Mixed Coniferous selection			Fir Selection			
1	2	3	4	5	6	7	8	9	
Commer-cial	Un-com-mercial	S. Total	Commer-cial	Un-com-mercial	S. Total	Commer-cial	Un-com-mercial	S. Total	G. Total
15.33	40.67	25.66	17.69	28.30	9.11	51.42	48.58	8.53	100.0

Source: A Digest of Statistics, 1982.

Note: The figures for commercial and non-commercial area percentages to their respective sub-totals and sub-total figures are percentages to total managed area in the State.

Geo-physical conditions of the State exhibit wide inter- and intra-regional variations which have consistently hampered the smooth functioning of forest management. Accordingly the State Forest Department had to adopt different systems of working in different regions. Some of the forests were given on long leases to big lessees who were expected to invest capital as the risk involved was less and profits assured. However, under this system the big forest lessees exploited the forests ruthlessly.

Consequently a new system, i.e. *sales by royalty* was introduced in order to remove the anomalies of the earlier system. It envisaged marking of trees in advance in accordance with the Plan prescriptions. This system could not also function properly due to administrative malpractices and numerous other bottlenecks. Along with these methods the State Forest Department established a lumbering project, which has now been changed into a full-fledged corporation, aiming to achieve the under-mentioned objectives:

(i) Replacement of orthodox methods of timber harvesting by introducing mechanical logging,
(ii) Optimal exploitation by introducing scientific methods of farm management,
(iii) Elimination of intermediaries and improvement in the conditions of labourers working in the forests, and
(iv) Removal of malpractices inherent in the agency system.

The departmental working of the forests is considered a step towards nationalization of forests. More recently, an ambitious project of reforestation called "Social Forestry" with the financial and technical help from the World Bank has been launched in the State. This project envisages to bring 44,000 hectares under various categories of plantation and plant 110 million new plants over a period of five years. The main thrust of plantation would be to cover strip lands, wet lands, degraded forests, village waste lands and to build up new forest nurseries for cultivating seedlings to be distributed to farmers free.

From the preceding discussions the following inferences can safely be drawn:

(a) Our forest policies and programmes have been formulated and initiated against the backdrop of reckless felling of trees. Conservation has been the main thrust of these programmes and policies. The vast potentialities of our forests (which could serve as a definite factor of economic development) have not been tapped fully.
(b) Without underestimating the pioneering work done by the State Forest Department since 1891 (when organized forestry took birth in the State under the expert guidance of McDonnell) the fact remains that the State's forests have been denuded. There

has been too much poaching and indiscriminate felling of trees. Forest wealth has either been overexploited or unexploited (both are negative trends). While precise statistical evidence may not be available to supplement these assertions the forest wealth of the State has undoubtedly been frittered away due to the collaborative activities of unscrupulous forest lessees and corrupt officials; cattle grazing; blasting for road making; ravages of weather; fires; military operations; conversion of forest land into agricultural land; lack of adequate and proper methods of forest farming; reforestation, management, harvesting and marketing; insufficient inputs and other administrative infrastructure; lack of alternative cheap sources of house building/packing material; and lack of political will to manage our forests on scientific lines. Vast tracts of land in Kandi Regions of Jammu have been denuded. Pir Panchal Range, the longest mountain range of the State separating Jammu from Kashmir, has an alpine blank in half of its area of 1,424 sq.km. The vast high sandy altitude of Ladakh is almost barren.

(c) There has definitely been a positive net revenue from forests. Expenditure on forests has also increased over the years. But a positive net revenue is no consolation when the opportunity cost of land is not taken into consideration. While there are instances of reckless deforestation conversion obsession has resulted in mature trees' dying and rotting.

Planning

In the light of the above conclusions planning for forestry in general and for forest fuel energy in particular needs to be revamped precisely because of the following reasons:

1. Wood will continue to dominate the consumption pattern of energy in the State for a long time to come.
2. Alternative supplies of energy will continue to remain uncertain or in some cases even unjustified.
3. Forests will continue to serve as a necessary adjunct to a number of industries.
4. Alternative uses of wood are likely to increase with the application of scientific methods particularly forest genetics.

Demand for wood from the industrial sector will also increase over the years.

5. Externalities associated with forestry are extremely necessary to tackle the problems of eco-development in the State.

Approaches to Planning

A fundamental promise in formulating a sectoral plan for forests is that it should be closely in consonance with the Plan objectives both in terms of time and space. For instance, for the transformation of State economy ,promotion of a viable rural economy is a must. If the rural sector is accepted as a crucial ingredient of development process then planning for meeting the basic needs of rural people such as food, fibre, fuel, shelter and employment from local sources is important. Further, if self-sufficiency or self-reliance is taken as a guiding principle of planning for the basic needs then local resources are to be preferred to the borrowed ones. In this respect trees cannot give only fuel, fibre and shelter but even jobs. Thus planning for fuelwood has to be a part of the planning for forestry as a whole. Planning for forestry has to be a part of the planning for other related sectors such as animal husbandry, agriculture, fisheries, etc., which in turn has to be a part of planning for the economy as a whole. At the same time each sector has its special characteristics and these are to be analysed and planned for.

Similarly the conceptual framework needed for the study of pollution control is the same as one required for fisheries, groundwater basins, forestry and so on. Therefore, concepts, tools and techniques of planning such as linear programming, social cost-benefit analysis, project evaluation, application of production functions, etc., cannot vary from sector to sector as far as their application is concerned.

Thus our approach to planning for forest fuel energy has to be a *Systems Approach* and the main thrust of this approach should be:

(a) To meet partially or in toto the long-term domestic fuel needs of the State particularly needs of the poorer sections of society,
(b) To supply fuel for industrial use and export wherever comparative advantages could be shown to exist,
(c) To devise ways and means whereby more wood per capita could be consumed without cutting additional trees,
(d) To manage forests on scientific lines,
(e) Encourage private sector to invest in fuelwood plantations, and

(f) Improve the methods of pricing, transportation, marketing and distribution of fuelwood.

Coverage

The planning for forest energy should cover all the regions of the State and sub-regions thereof. People's involvement is to be ensured at all levels. The basic unit of planning should be a block.

Demand Projections

The demand for fuel is the function of climate, cooking habits of people, family size, availability of alternative sources of energy and social attitudes towards wood. The per capita fuelwood requirements of each climatic zone and each section of society should be calculated by conducting comprehensive and regular surveys by private and public agencies. Demand forecasts projected from these surveys should take into consideration various factors expected to affect the future consumption of a particular fuel such as shifts in the level and structure of consumption pattern resulting from either improvement in the standard of living or change in the indices of measuring the standard of life. Demand projects are also necessary to ensure that the long-term domestic needs are not jeopardized by the industries established to utilize as raw material the wood which is needed for fuel and charcoal.

Resource Balance-Sheet

In order to ensure a regular supply of fuelwood the amount of wood available from forests and also the area which could be made available for fuel plantation is to be assessed properly. Land requirements of other sectors of the economy are to be projected so that reforestation and forestation programmes can be accommodated in the framework of the whole economy. The proportion of land under forests is not a reliable and adequate indicator of the optimum size of forest area unless comparative advantages of exporting timber, minor forest products and home demand for fuelwood are taken into consideration. It is a management problem and is to be based on reconciliation of conflicting alternatives which are often affected by restrictions. Linear programming can help to resolve the complex problems associated with the best available permutations and combinations of activities.

The technique will also take care of variations of the optimum which may provide the crucial guidance on the direction which the management should take. If maximization of the provision of forest fuel is set as an objective function, after proper demand forecasting, the factors which are relevant to the satisfaction of the demand must be enumerated and assessed. The sustained supply of forest energy depends upon the incremental product of trees. The existing forest coverage may be capable of yielding this sustained supply of energy from thinnings, refining operations or from the residues of final harvesting. The amount available should be estimated for each forest division/tract and specieswise along with the effective potential demand of competing alternatives. The difference between expected supply and demand for fuelwood will have to be provided from alternative sources. Studying of growth functions of trees and regular forest farm-management surveys will facilitate the whole process. Land for growing fuel has an opportunity cost which must be approximately accounted for while making planning decisions.

Selection of Suitable Projects

To maximize the objective function of fuelwood planning application of principles of project planning is extremely important.

Management and Administration

The existing management philosophy is the legacy of traditional conservation forestry practices and is not conducive to dynamic forest development approach. The administrative set-up is either incapable or overburdened to undertake fast and intensive development programmes. There is a lot of unnecessary centralization and overlapping of functions. This situation needs to be changed. The planning for forest energy should be carried under the supervision of a team comprising economists, soil chemists, agronomists, entomologists, experts of forest genetics and people trained in forest administration. A forest fuel production plan, which will be part of the plan for the forest sector, should be worked out by an economist with the help of these experts. Similarly new methods of forest farming, training, vintage, disease and insect control, fertilization, use of bio-mass machines for optimum harvesting, improving the methods of sowing, reforestation by scattering seedlings by helicopters, increased use of forest genetics to produce super-trees, growing of test-tube red-wood, saving energy by improving the insulating qualities of wood, extracting oil, natural gas, turpentine, tall oil from wood, and producing basic chemicals from wood such as phenol, estones, benzene, ammonia and formaldehydes will go a long way in helping us to use forests in accelerating economic development. Economics of forestry should also be introduced at undergraduate and graduate courses of studies. Para-statal organizations such as fuel corporations should be created which will deal with:

(a) Marketing, purchase and transportation of fuelwood.
(b) Grow its own fuelwood plantations and encourage households to cultivate woodlots for commercial purposes.

Pricing of Firewood

The present system of prices of fuelwood is extremely defective. Huge quantities of firewood are supplied to the civilians and army on subsidized rates resulting in waste of financial resources which could otherwise be utilized to create productive forest assets. Lack of adequate roads and quick methods of transport are other important bottlenecks which could be improved to bring down the cost of timber and wood extraction.

Conclusion

It can safely be concluded that if we learn the art of husbanding our trees, we can not only meet the challenges of energy but produce much more than what we possibly think of at present. Forestry is a working partnership between common sense and what T. M. Huxley called organized common sense. We might have been guided by common sense but definitely not by organized common sense in evolving the present forest management policies and programmes. Here we are reminded of Harnam Singh Pathania (Chief Conservator of Forests, Jammu and Kashmir, in the early fifties) refuting the chances of reckless felling and asserting before a researcher:

> Whenever systematic felling was being done forests were being replaced with new crop and regeneration added. One has to move out to see fine work done.

But unfortunately today when we go out instead of *fine work* we find degraded forests, multi-crore rosin scandals, crores of rupees outstanding against forest lessees and as many as 77 listed endangered species of mammals, reptiles, birds and insects. The forests, about which our grandfathers used to narrate us fairy tales of *Forest Kingdom* have been turned into agricultural lands or cricket fields, with the result that we neither play good cricket nor produce sufficient rice. Under the circumstances we can do just nothing but recall the touching words of Charles Dickens:

> It is the worst of times, it is the best of times,
> it is the age of foolishness, it is the age of wisdom,
> it is the winter of despair, it is the spring of hope,
> we have nothing before us, we have everything before us.

APPENDIX I

Methodology

With a view to having an idea about the trend in the movements in time series,* trend regression coefficients have been calculated by adopting the function:

$$Y_i = b_o \ b_i X_i$$

The values of the parameters have been estimated by using the two normal equations:

$$\varepsilon Y_i = Nb_o + b_i \varepsilon X_i$$

$$\varepsilon X_i Y_i = b_o \varepsilon X_i + b_1 \varepsilon X^2_i$$

which while solving simultaneously are reduced to**

$$b_1 = \frac{\varepsilon x_i y_i}{\varepsilon x^2_i}$$

where x_i and y_i are deviations from the respective means and

$$b_o = \overline{y} - b_1 \overline{x}$$

In order to measure the size of error and determine the degree of confidence in the validity of the estimates, the standard error test has been employed with the formula***

---

* Assuming that . . . the successive observations are statistically independent. . . . See Yamane Taro, *Statistics, An Introductory Analysis*, Harper International Edition, New York 1964, p. 865.

** Gujrati, D., *Basic Econometrics*, McGraw Hill International Book Company, 1979, pp. 41-43.

*** Koutsoyiannis, *Theory of Econometrics*, Second Edition, The Macmillan Press Ltd., London, p. 81.

$$SE_{(b_1)} = \frac{\varepsilon (Y-Yc)^2}{(n-2)\ \varepsilon x^2}$$

To test the statistical significance of the coefficients, 't' statistic has been employed in the usual way:

$$t_{b^1} = \frac{b_1}{SE\ (b_1)} \quad \text{with n--2 degrees of freedom}$$

To measure the growth overtime of the trend variables constant growth model has been employed which in its normal form can be written as:

$$Y = ab^T$$

Taking logs to both sides, we get:

$$\text{Log } Y = \text{Log } a + T \text{ Log } b$$

The coefficients have been calculated with the help of two normal equations:

$$\varepsilon \text{ Log } Y = N \log a + \text{Log } b\ \varepsilon T$$

$$\varepsilon\ T \text{ Log } Y = \text{Log } a\ \varepsilon T + \text{Log } b\ \varepsilon T^2$$

A compound growth rate calculated as (b-1) 100 has been worked out showing the percentage increase/decrease in the dependent variable with a unit change in independent variable.

Sources of Data

Time series data on different aspects of the State forests have been compiled and tabulated from the following published sources:

1. *A Digest of Forest Statistics, Department of Forest*, 1982,
2. *Digests of Statistics,* Directorate of Statistics, 1965-66, 1968-69, 1973-74, 1975-76 and 1981-82,

3. *Indicators of Economic Development of Jammu and Kashmir*, Directorate of Economics and Statistics, 1984,
4. *Pilot Survey Report*, Directorate of Economics and Statistics, Wood Balance Unit, August 1983-July 1984.

Conceptual Schemes

The definitions of different concepts used by the abovementioned agencies/departments have been taken as such.

Coverage

Data of (i) Area, (ii) Outturn of timber, (iii) Firewood, (iv) Major forest products, (v) Minor forest products and (vi) Revenue have been collected from 1951-52 to 1983-84 while the data on net surplus and expenditure on forests have been compiled from 1950-51 to 1981-82.

The analysis has been made planwise, decadal and on overall basis.

REFERENCES

Bhat, M.S. (1977) "Kashmir Forests: Uses and Prospects" *Mainstream*, Vol. XV, No. 52, Aug. 27.

Chandras, G.S. and Muranjan, S.W. (1968) "Intensification of Forestry Production through "Area Selection" *Artha Vijnana*, Vol.10, No.2, June.

Das Gupta, P. (1982) *The Control of Resources*, Basil Blackwell Publishers Ltd.

Dorfman, R. and Dorfman, N.S. (eds) (1972) *Economics of the Environment: Selected Readings*, W. W. Norton, New York.

Earl, D.E. (1975) *Forest Energy and Economic Development*, Clarendon Press, Oxford.

Faustmann, G. (1949) "On the Determination of the Value which forest land and immature stands possess for forestry" reprinted in English in *Oxford Institute Papers* (1968), 42.

Freeman, A.M. (1979) *The Benefits of Environmental Improvements: Theory and Practice*, Johns Hopkins University Press, Baltimore.

Freeman, A.M. Haveman, R.H. and Keneese, A.V. (1973) *The Economics of Environmental Policy*, John Wiley, New York.

Futehally, Z. (1985) "The Importance of Trees" *The Indian Express*, May 4.

Hartwich, J. (1980) "Principles of Natural Resource Use", Deptt. of Economics, Queens University, Ontario (Mimeo).

Joseph, K.J. (1978) "Planning for Forestry Sector" *Economic Times*, May 25, 26 and 27.

Mitra, T. and Wan, H. (1981) "On the Foustman Solution to the Forest Management Problem", Deptt. of Economics, Cornell University (Mimeo).

Niehaus, F. (1980) "Developing Criteria to Compare the Safety of Energy Systems", Angewandte System analyse Band, 1.

Person, R.(1974) *World Forest Resources*, Royal College of Forestry, Stockholm.

Samuelson, P.A. (1976) "Economics of Forestry in an Evolving Society", *Economic Inquiry*, 14.

Seminar June 1985, pp. 12-16.

Smith, V.L. (1968)" Economics of Production from Natural Resources" *American Economic Review*, 58(3).

Sommer, A. (1976) "Attempt at an Assessment of the World's Tropical Forests", *Unasylva*, 28.

Svendsrud, A. (ed) (1969) "Forest Economics", Universitats-forlagest, Oslo.

Thapar, S.D. (1975) *India's Forest Resources*, Birla Institute of Scientific Research, 1975.

UNESCO (1978) *Tropical Forest Eco-Systems,* ISBN UNESCO, Paris.

Wiegner, K.K. (1980)"Forests: Green Gold", *Economic Impact*: Number thirty 1980/2.

Reports

Annual Administration Report on Forest Administration in Jammu and Kashmir for the year 1892 to 1951-52.

Notes on a tour in the forest of Jammu and Kashmir State in 1905 by S. Eardly Wilnot, Inspector-General of Forests, to the Government of India.

Note on Tour of Inspection in some of the forests of State (1920) by Sir George Mart (K.B.S.C.I.E), Inspector-General of Forests, to the Government of India.

Note on the State Forests by Sir Alexander Rodger, C.B.E, Inspector-General of Forests, to Govt. of India (1930).

Forest Enquiry Committee's Report of 1924.

Imperial Gazetteer of India, vol. XV, by Hanery Forwede.

India's Year Book, 1953.

Annual Administration Report of the Forest Administration for the year 1952-53 to 1954-55.

Pilot Survey — Consumption of Firewood and its by-products together with other fuels, Directorate of Economics and Statistics, Planning and Development Deptt., Wood Balance Unit, Aug. 1983-July 1984.

Social Forestry Manual, published by Social Forestry Organisation, J and K Forest Deptt. April 1985.

Social Forestry in the Service of Society, issued by the Directorate of Social Forestry Project, Jammu and Kashmir.

Digest of Statistics, Directorate of Economics and Statistics, Planning and Development Deptt. 1965-66; 1968-69; 1973-74; 1975-76; and 1981-82

Indicators of Economic Development of Jammu and Kashmir, Directorate of Economics and Statistics, 1984.

A Digest of Forest Statistics, Jammu and Kashmir Forest Record No. 1 (Fourth Edition), Srinagar, 1982.

SESSION III

PROSPECTS OF NUCLEAR POWER IN J&K STATE

Chairman: Dr. M.R. Srinivasan, Chairman, Nuclear Power Board, Bhaba Atomic Research Centre

---

ELEVEN

NUCLEAR REACTORS AND FUEL CYCLES

P.K. IYENGAR

INTRODUCTION

The Sun has been the primary source of energy needed by man. It provides him with vegetation to meet his requirements of food and heating. The vegetation also supports a variety of fauna and some of these provide man with food and motive power. Over millions of years, large tracts of animal and vegetable kingdom got buried deep under the soil and were converted to oil and coal by nature. Man discovered these sources a few hundred years ago and has become increasingly dependent upon them for meeting his energy needs. Availability of these rapidly exploitable sources has contributed to his progress in a large measure. The Sun also provides man with energy by ensuring a perpetual flow of water in the streams and rivers which can be harnessed to produce hydel power. Energy has become the food for new technological devices, a consistent stream of hope for modern man and a prerequisite for his survival.

Most of the energy used by man comes *via* chemical reactions. For example, in burning of coal the atoms of carbon react with oxygen to form carbon dioxide, burning of each atom of carbon giving about two electron volts of energy. But in the Sun, which is the primary source, a nuclear reaction involving fusion of light elements like hydrogen, deuterium, tritium, etc., to form heavier elements is taking place and energy released per conversion is 14 million electron volts. It is thus seen that nuclear reactions can lead to production of much larger quantities of energy. Efforts to bring about a controlled fusion reaction are in progress in many countries in the world but the feasibility of producing net energy output is yet to be demonstrated. Uncontrolled fusion reaction takes place in a hydrogen bomb and all of us know the tremendous quantity of energy released in such explosions. A controlled nuclear reaction which has been harnessed by man is the fission reaction which involves the splitting of heavy elements like uranium. When we talk of nuclear energy today, we are talking about the reactors based on the fission reactions. Here I would like to discuss some of the fundamentals for the production of nuclear power, the type of reactor systems which are presently in use, the nuclear fuel cycles for these reactors and the safety of nuclear power.

The Fission Reaction

Two naturally occurring elements which are of importance in nuclear power production are uranium and thorium. Uranium has two isotopes, uranium-235 and uranium-238. A typical fission reaction involving the splitting of uranium-235 nucleus is shown in Fig. 11.1. The breakup is accompanied by the release of 200 million electron volts of energy. Fission of one gram of uranium-235 produces one MWD of thermal energy which would normally require burning of 2.5 tonnes of coal. Two to three neutrons having energy in the range of 0.1 MeV to 10 MeV are produced per fission event which play an important role in the production of nuclear energy (Fig. 11.2). Neutrons of all energies can cause fission in uranium-235 but in the case of uranium-238 fission reaction has reasonable probability only when neutron energy is greater than 1.4 MeV. Uranium-235, because of its ready fissionability, is called fissile isotope and it happens to be the only naturally occurring fissile isotope.

Generation of nuclear power requires propagation of fission reaction through neutrons. If sufficient care is taken to minimize the escape of neutrons, or their capture by other materials, so that a sufficient number interact with uranium-235, a fission chain reaction can be sustained. Unfortunately the concentration of uranium-235 in natural uranium is only 0.72% (1 atom out of 140 atoms) and fission chain reaction cannot be sustained by high energy neutrons produced in the fission reaction. Slowing down (moderation or thermalization) of fission neutrons from their average energy of 2.0 MeV to 0.024 MeV increases the probability of fission reaction several hundred times and chain reaction can be sustained with natural uranium as fuel if a suitable moderator is chosen to slow down the neutrons. Materials which are good moderators include water (light water in nuclear terminology), heavy water (water containing deuterium, the heavier isotope of hydrogen) and graphite, but out of these only the latter two are suitable for use with natural uranium. The first atomic pile which demonstrated the sustenance of fission chain reaction was operated in 1942 with uranium metal as fuel and graphite as moderator. Light water contains hydrogen nuclei which absorb neutrons. Enrichment of uranium to increase its uranium-235 content is required if light water is to be used as a moderator. Use of a moderator is however not essential for sustaining the fission chain reaction if the concentration of fissile isotopes is increased 10 to 20 times above the natural level the chain reaction can be sustained even with fast neutrons.

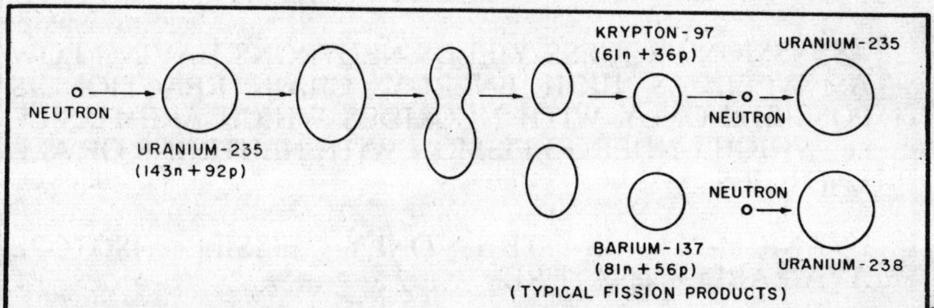

$${}^{1}_{0}n + {}^{235}_{92}U \longrightarrow {}^{137}_{56}Ba + {}^{97}_{36}Kr + 2\,{}^{1}_{0}n + 200\,MeV$$

FISSION OF ONE GRAM OF ^{235}U GIVES ONE MEGAWATT DAY OF THERMAL ENERGY = ENERGY RELEASED BY BURNING 2.5 TONS OF COAL

Fig. 11.1: Nuclear Energy.

Nuclear Steam Supply System

Even though the first atomic pile which demonstrated the possibility of harnessing fission energy was built in 1942, the first reactor to actually demonstrate power production was built in 1951 (EBR-I in USA) and the first commercial plant came on stream only in 1956. (Calder Hall, U.K.) This time was required to develop the engineering side of the system. It is not enough if the chain reaction is sustained. One must have the right temperature and power density inside the core of the reactor so that the coolant gets properly heated along with uniform cooling of core and high pressure steam can be generated for running a turbine. A schematic view of a reactor of the type we have at Tarapur, namely, the Boiling Water Reactor (BWR) is shown in Fig. 11.3. Historically this type of reactor was initially designed as a small reactor (40 MWe) for propulsion of sub-marines. This design was chosen because a very compact reactor could be made which could readily be fitted inside a sub-marine. This type of reactor uses enriched uranium, i.e., uranium in which uranium-235 content has been

NATURAL URANIUM	→	0.7% URANIUM-235 + 99.3% URANIUM-238
URANIUM-235	→	UNDERGOES FISSION WITH NEUTRONS OF ANY ENERGY
URANIUM-238	→	UNDERGOES FISSION ONLY WITH NEUTRONS OF HIGH ENERGY (> 1.4 MeV)

AS FISSION PROCESS YIELDS NEUTRONS HAVING LOW AS WELL AS HIGH ENERGY, CHAIN REACTION IS POSSIBLE ONLY WITH NUCLIDES WHICH ARE FISSILE i.e. WHICH UNDERGO FISSION WITH NEUTRONS OF ALL ENERGIES.

URANIUM-235 IS THE ONLY FISSILE ISOTOPE AVAILABLE IN NATURE.

Fig. 11.2: Fissile Elements — 1.

raised by artificial means, and the whole core can therefore be just one metre in diameter. Ordinary water is used as the moderator and coolant and the steam is produced within the reactor itself. In the upper portion of the reactor the steam is separated from water and goes to turbine generator for the production of electricity. The reactor is housed in a pressure vessel that can readily withstand pressures much higher than those generated by steam/water. The reactor is housed in a strong containment building which isolates the nuclear steam supply system from the conventional parts like turbine and auxiliaries. From the very beginning the nuclear scientist has been very conscious of the hazards due to the radiation-emitting fission products produced during fission reaction and has taken abundant precautions to ensure that these are not released to the environment. The fuel is first contained in a strong metallic tube made out of a special alloy of zirconium. This fuel is housed inside the pressure vessel and the pressure vessel itself is housed in a containment structure. Thus there are three barriers to the release of radioactive fission products. Further, the containment structure has special devices to condense vapour and also high efficiency filtration systems for removing radioactivity should any vapour leak out of the containment structure. The capacities of this type of reactors have been gradually increasing over the years and presently reactors having 1,200 MWe capacity each are quite common. What is involved in increasing the capacity is the size of the pressure vessel, the kind of pumps to be used, the turbine generator, etc.

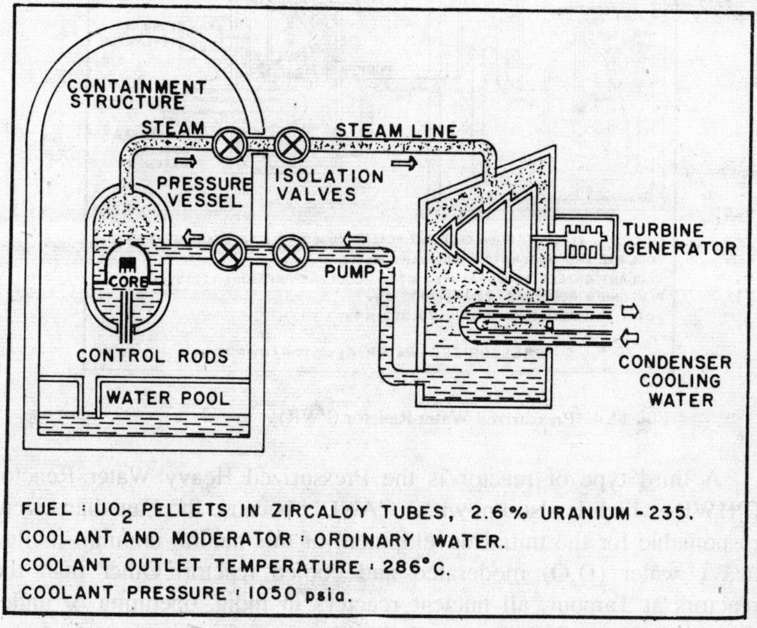

Fig. 11.3: Boiling Water Reactor (BWR).

Perhaps Japan has some of the most modern boiling water reactors operating today. The aggregate capacity of this type of reactor in the world is presently about 50,000 MWe.

Another reactor cooled and moderated by light water is the pressurized water reactor (PWR) shown in Fig. 11.4. In this reactor the pressure of water used for cooling the reactor is kept high so that it does not boil even when the temperature reaches around 300°C. This pressurized water at high temperature is used for generating steam in a steam generator and is returned to the reactor. All features other than this are essentially identical to the BWR reactors. This type of reactor is most popular in the world and presently the aggregate capacity of PWRs operating in the world is about 117,500 MWe.

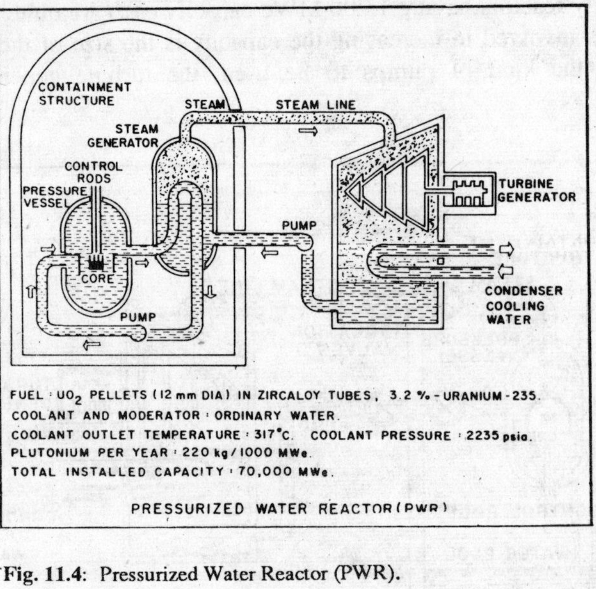

FUEL : UO_2 PELLETS (12 mm DIA) IN ZIRCALOY TUBES, 3.2 %-URANIUM-235.
COOLANT AND MODERATOR : ORDINARY WATER.
COOLANT OUTLET TEMPERATURE : 317°C. COOLANT PRESSURE : 2235 psia.
PLUTONIUM PER YEAR : 220 kg/1000 MWe.
TOTAL INSTALLED CAPACITY : 70,000 MWe.

PRESSURIZED WATER REACTOR (PWR)

Fig. 11.4: Pressurized Water Reactor (PWR).

A third type of reactor is the Pressurized Heavy Water Reactor (PHWR) which is also known as CANDU because the Canadians were responsible for the initial development of this natural uranium-fuelled heavy water (D_2O) moderated and cooled reactor. Other than the reactors at Tarapur, all nuclear reactors in India, operating or under construction, are PHWR type. India has opted for this type of reactor because of many reasons including the fact that we wanted to be self-reliant in this field and the heavy water in the desired quantities could be readily produced by special chemical enrichment methods from ordinary water and natural uranium could be used as fuel without any need for enrichment.

In this reactor (Fig. 11.5) various components are housed inside a vessel called calandria, which is not a pressure vessel. It is simply a horizontal cylindrical vessel containing heavy water in which several pressure tubes are inserted for containing the fuel bundles. These tubes are also made out of special zirconium alloy and are put in such a way that fresh fuel bundles can be inserted from one side and used up fuel bundles can be taken out from the other side. Thus fuel in this type of reactor can be replenished even while the reactor is operating whereas BWR and PWR have to be shut down for refuelling. On-line refuelling is essential for PHWRs because the fuel contains only a small amount of uranium-235 which gets readily burnt. The fuel in the pressure tubes is cooled by pressurized heavy water which in turn is used for the generation of steam in a steam generator.

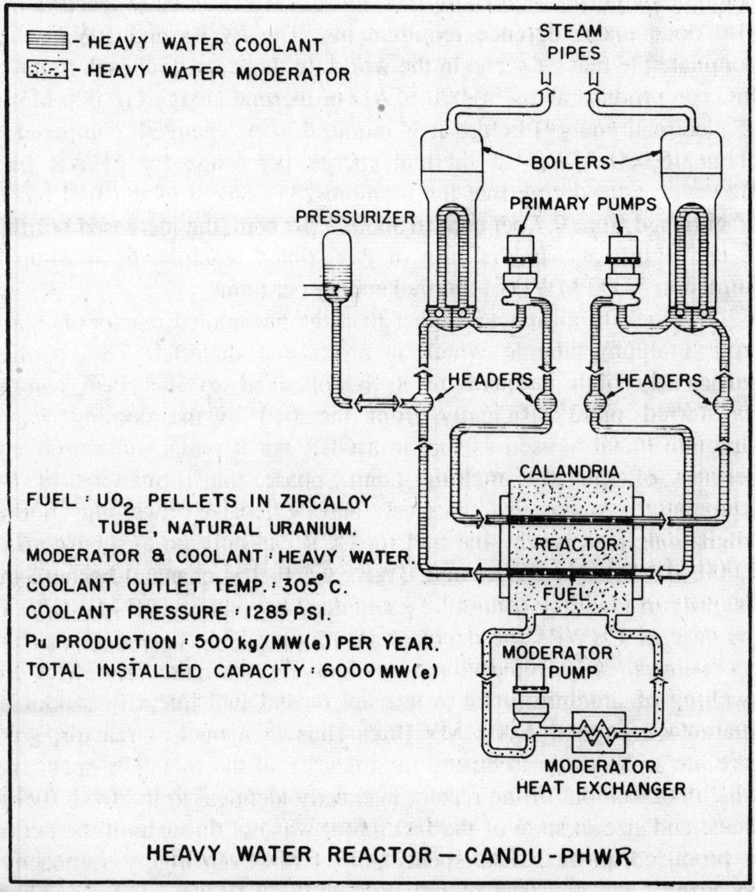

Fig. 11.5: Heavy Water Reactor — Candu-PHWR.

Historically, gas cooled reactors (GCRs) pioneered by Britain and France were the first to become commercial. Calder Hall plant in Britain became operational in 1956. Presently Britain is building advanced gas cooled reactors. I am emphasizing these things to point out that each country chooses a technology and then carries on with it because they have developed the necessary expertise and would like to make it a success and not depend upon borrowed technology. The BWR and PWR concepts emerged out of the development of reactors for sub-marines in the U.S.A. Thanks to their defence requirements, they had an enormous capacity to make enriched uranium and therefore they could afford to make enriched fuel available for commercial purposes also. Further, most of R&D on these reactors was also done under defence requirements. The PWRs and BWRs thus dominate the reactor scene in the world. In these reactors each tonne of fuel can produce about 30,000 MWD of thermal energy (10,000 MWD of electrical energy) before it is required to be changed, compared to about 10,000 MWD of thermal energy per tonne for PHWR fuel. However, considering that the uranium-235 content of the fuel has to be enriched from 0.7 per cent to about 3 per cent, the increased burn-up is not significant. In the case of gas cooled reactors the burn-up is limited to 5,000 MWD of thermal energy per tonne.

The fuel for all reactors other than the gas-cooled reactors is made from uranium dioxide which is a ceramic material. This permits sufficiently high temperature to be obtained so that heat can be transferred more efficiently from the fuel to the cooling water. Uranium metal is used as fuel in a GCR but it poses some problems because of the low melting point, phase transformations in the temperature region of interest and extensive swelling during irradiation. That is why the fuel for GCR can only go to a burn-up of 5,000 MWD/te. In PWRs and BWRs the fuel is changed because the integrity of cladding cannot be guaranteed beyond 30,000 MWD/te. In the case of PHWRs the decrease in uranium-235 content of the fuel necessitates early replacement of fuel. In the case of GCRs the swelling of uranium metal is too severe and fuel integrity cannot be guaranteed beyond 5,000 MWD/te. Thus in a nuclear reactor, great care has to be taken to ensure the integrity of the fuel. The spent fuel which is taken out of the reactor is exactly identical to the fresh fuel in shape and size in spite of the fact that it was hot throughout the period it produced power. The spent fuel is however highly radioactive because of the presence of radiation-emitting fission products. These radioactive products are safely contained inside the fuel cladding and do not escape to the environment. However, this fuel has to be handled with remote handling tools to protect personnel from radiation. The quantity of fuel required for nuclear reactors is however very small, e.g. each reactor at Tarapur requires only 15 tonnes of fuel each year compared to more than 15 lakh tonnes for a similar sized coal fired plant.

The Concept of Breeding

Conversion and Breeding

Uranium-238, the major isotope of uranium, also has an important role as it can absorb some of the neutrons released in the fission reaction to form uranium-239 which decays *via* neptunium to fissile isotope plutonium-239 (Figs. 11.5 & 11.6). Another fissile isotope of plutonium which is formed in a nuclear reactor is plutonium-241 and these two fissile isotopes constitute 70 to 90 per cent of the plutonium produced in the nuclear reactors.

Similarly, there is another element thorium-232, which is quite abundant in our country, which does not by itself undergo fission in a reactor but can be converted to uranium-233 which is 100 per cent fissile. Thus even though nature provided us only with uranium-235, we can produce plutonium-239 and uranium-233 in sufficient quantities to sustain a nuclear power programme.

The consumption of fissile uranium-235 is always accompanied by the production of fissile plutonium (or uranium-233 if thorium is present) but in all the reactors which use moderators this process is down-hill, i.e. the fissile material produced is significantly less than the fissile material consumed. The ratio of these two quantities, which is called the conversion ratio, is shown for various types of reactors in Fig. 11.7. It is seen that this ratio has a value of 0.55 for light water reactors and 0.71 for heavy water reactors. When conversion ratio exceeds unity it is called the breeding ratio and the reactors in which this is possible are called breeder reactors. The secret of breeding can be better understood from Fig. 11.8 which depicts the number of neutrons produced per fissile atom consumed as a function of neutron energy. It is seen from this figure that uranium-233 gives the maximum number of neutrons up to the neutron energy of 0.1 MeV beyond which the neutron yield from plutonium-239 increases very sharply. Thus if we use plutonium as fuel and neutron energy is kept substantially above 0.1 MeV, 2.5 to 3 neutrons are released out of which one is required for sustaining the chain reaction and the balance are available for absorption in uranium-238 for its conversion to plutonium. Further, when the neutron energy is high, uranium-238 also undergoes fission thereby contributing to additional neutrons for breeding. A breeding ratio of approximately 1.2 is achievable with the existing technology of fast reactors which use mixed oxide of uranium and plutonium as fuel. Some neutrons are lost due to absorption by structural materials and some leak out of the reactor. A breeding ratio of 1.3 or more may be accomplished in advanced fast reactors. The concept of breeder can be understood more readily from Fig. 11.9. A 1,000 MWe breeder reactor would use about 3 tonnes of plutonium and 12 tonnes of uranium-238 as fuel. After producing power for one year, in which about 0.8 tonnes of plutonium and 0.2 tonnes of uranium would undergo fission, 3.2 tonnes of plutonium, 10.8 tonnes of uranium-238 and 1 tonne of fission products are left in the spent fuel. Simultaneous production of electricity as well as excess fuel is a unique feature of nuclear energy.

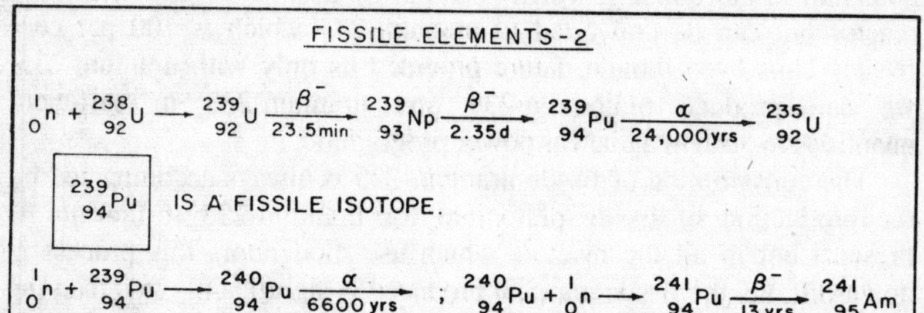

Fig. 11.6: Fissile Elements — 2.

CONVERSION RATIO = $\frac{\text{FISSILE ATOMS PRODUCED}}{\text{FISSILE ATOMS DESTROYED}}$

LIGHT WATER REACTORS : 0.55
HEAVY WATER REACTORS : 0.71
CONVERSION RATIO > 1 IS CALLED BREEDING RATIO

Fig. 11.7

Fast Breeder Reactor

A schematic view of the Liquid Metal Cooled Fast Breeder Reactor which is under development in various countries is shown in Fig. 11.10. This reactor uses mixed oxides of uranium and plutonium as fuel with plutonium concentration ranging between 15 and 25 per cent. It is small in size, uses no moderator and the core is surrounded by uranium-238 or thorium to achieve breeding. Liquid sodium is used as a coolant to achieve a higher cooling rate since temperature in the core is high. Since liquid sodium reacts violently with water, sodium used for cooling the reactor core is not directly used for generating steam to ensure safety of the reactor; secondary sodium loop is used for this purpose. The status of the fast breeder reactor development in various countries is shown in Fig. 11.11. The USSR has a strong commitment to fast breeders and at present has the largest operating reactor of this

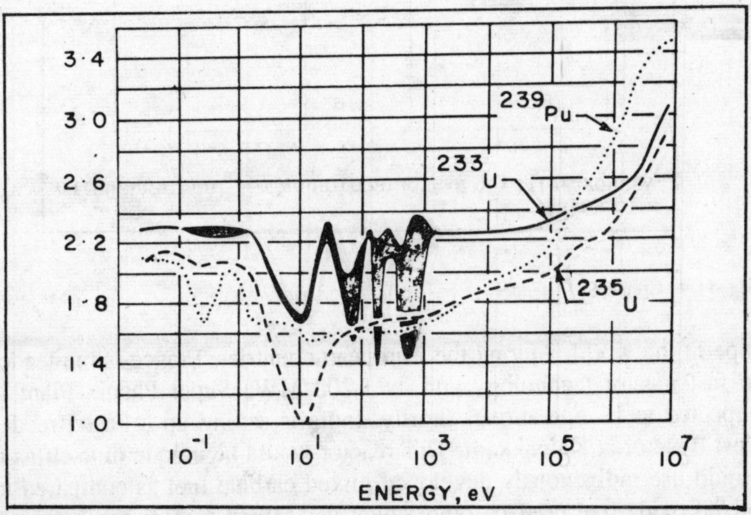

Fig. 11.8: η Number of Neutrons Released Per Neutron Absorbed as a Function of Neutron Energy.

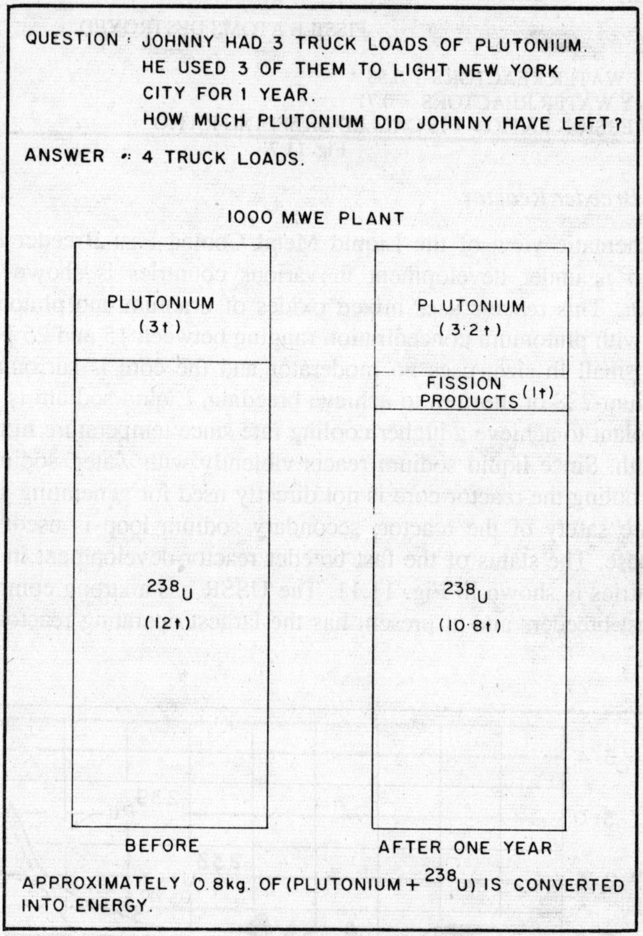

Fig. 11.9: Concept of Breeder.

type in the world. Among the European countries, France is the leader in fast reactor technology and its 1,200 MWe Super Phenix Plant is expected to be operational shortly. India is setting up a Fast Breeder Test Reactor at Kalpakkam. This reactor would be unique in itself as it would use indigenously developed mixed carbide fuel as compared to mixed oxide fuel used by other countries. Mixed carbide has a greater potential for breeding and is therefore an advanced fuel. The advantage of going in for fast reactors is that one can be quite independent of the constraints of natural uranium required for a power programme. Once you have plutonium, you keep on converting more depleted uranium or thorium into fissile material, so that the availability of sufficient amount of natural uranium is no longer a constraint. We are also in the process of designing a 500 MWe fast breeder reactor which would go into operation in 1995-96.

Nuclear Fuel Cycles

Now we turn our attention to the Nuclear Fuel Cycles. The fuel cycle for the present-day reactor is essentially based on uranium and is called Uranium Fuel Cycle. For fast reactor we may either have a Plutonium Fuel Cycle or a Thorium Fuel Cycle.

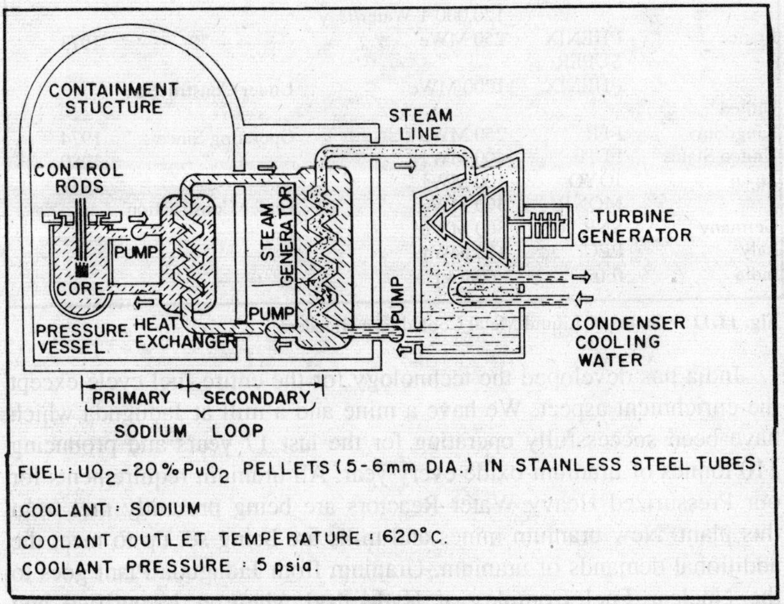

Fig. 11.10: Liquit Metal Fast Breeder Reactor (LMFBR).

Uranium Fuel Cycle

The fuel cycle for reactors using natural or slightly enriched uranium is shown in Fig. 11.12. The cycle starts with the mining and recovery of uranium and refining it to the nuclear grade. Refined uranium is directly sent to fuel fabrication plant for use in reactors like PHWRs.

For boiling water reactors (like Tarapur reactors) or pressurized water reactors, refined uranium is sent to an enrichment plant for increasing its uranium-235 content. The fabricated fuel goes to the reactor and is used for the production of electricity. After attaining the desired burn-up the fuel is taken out and transported to a reprocessing plant for the recovery of plutonium and depleted uranium. The radioactive fission products waste is immobilized and taken for safe storage.

Country	Reactor	Power	Status	
USSR	BN 600	600 MWe	Operating Since	1980
	BN 350	150 MWe + Desalination of 120,000 T Water/Day	——" ——	1973
France	PHENIX	250 MWe	——" ——	1973
	SUPER PHENIX	1200 MWe	Under Construction	
United Kingdom	PFR	250 MWe	Operating Since	1974
United States	FFTF	400 MW†	——" ——	1980
Japan	JOYO	100 MW†	——" ——	1977
	MONJU	300 MWe	Under Construction	
Germany	SNR 300	300 MWe	——" ——	
Italy	PEC	188 MW†	——" ——	
India	fbtr	60mw†	——" ——	

Fig. 11.11: Status of Liquid Metal Cooled Fast Breeders.

India has developed the technology for the entire fuel cycle except the enrichment aspect. We have a mine and a mill at Jaduguda which have been successfully operating for the last 17 years and producing 110 tonnes of uranium oxide every year. All uranium requirements for our Pressurized Heavy Water Reactors are being presently met from this plant. New uranium mines and mills are being set up to meet the additional demands of uranium. Uranium from Jaduguda Plant goes to the Nuclear Fuel Complex at Hyderabad where it is purified and converted to uranium dioxide pellets. Zircaloy tubes required for hermetically sealing these pellets are also fabricated at this plant starting from zircon sand obtained from beaches at Kerala. Pressure tubes made out of zircaloy and many other special components required for the fabrication of a reactor are manufactured at this plant.

India also has developed the know-how for design, construction and operation of the PHWRs and the first totally indigenous plant is successfully operating at Kalpakkam. The spent fuel from PHWRs is transported in specially designed containers to the reprocessing plant at Tarapur. At the reprocessing plant the fuel element is broken into small pieces and the fuel dissolved in nitric acid for the recovery of plutonium. Nuclear fuel reprocessing is one of the most challenging operations in chemical engineering. Although many of the basic steps are common to those in a conventional hydrometallurgical industry, the presence of large amounts of highly radioactive fission products dictates that the complete scheme of operations is carried out remotely. The radiation also makes any maintenance extremely difficult or sometimes even impossible and therefore exceptionally high standards of design, construction and installation are essential. Most of the components and equipment have to be ensured for life-time reliability. To overcome the problem of malfunction in various systems, some of these are duplicated or even multiplicated which makes the plant highly complex. You would get an idea of this complexity by an internal view of Tarapur plant shown in Plate 11.1. Uranium and plutonium are required to be separated from a very large number of fission product elements and the purity requirements surpass even

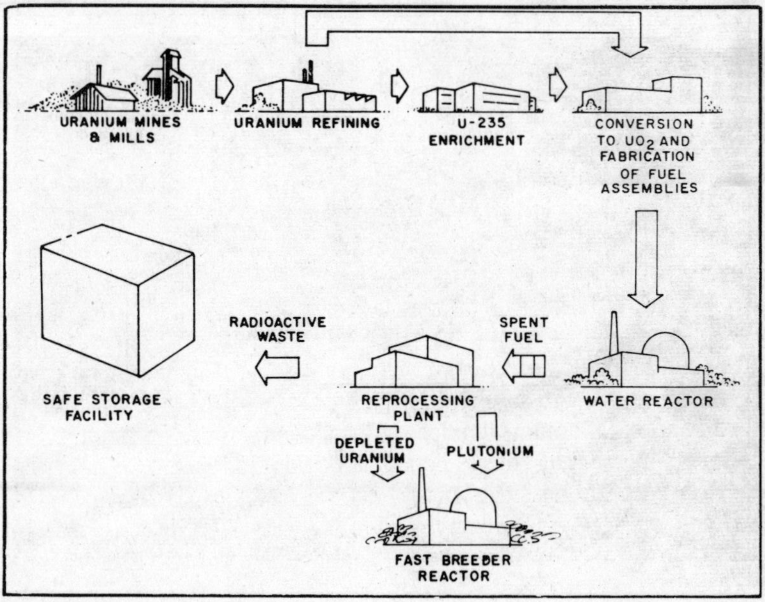

Fig. 11.12: Uranium Fuel Cycle.

Plate 11.1: Power Reactor Fuel Reprocessing Plant Tarapur — A Typical View of the Process Cell.

Plate 11.2: Facility for Immobilization of Reactive Waste.

those in the pharmaceutical industry. Further, plutonium is a highly valuable product and it is essential to ensure its near complete recovery. India is one of the few countries possessing the reprocessing technology and its plant at Tarapur has been successfully reprocessing fuel from reactors at Rajasthan. As fission products are exposed for the first time in the reprocessing plant great care is taken to see that they are not released to the environment either through the gaseous or through the liquid effluents. Special gas treatment systems are installed in a reprocessing plant and the gases sent out are continuously monitored for their radioactive content.

The waste stream containing the fission products has to be carefully treated because it contains essentially all the radioactivity. The quantities are not large since the weight of uranium fissioned is only about 10 kg per tonne of spent fuel. The problem of this waste is quite often highlighted in the newspapers. However, we have developed the technology to safely immobilize this waste in a glass matrix and Plate 11.2 shows an inside view of a cell used for this purpose. The quantities of glass generated are quite small. For example, our programme of 10,000 MWe would produce only 40 m^3 of glass (with immobilized fission products) up to the year 2000. This small quantity of glass can be easily stored safely for about 25 years and disposed off in geological formations, hard rocks or salt mines. A facility for safe storage up to 25 years is already being set up at Tarapur.

Plutonium Fuel Cycle

The first stage of India's nuclear power programme is based on the use of natural uranium in PHWRs. Adequate uranium resources have been identified to meet the life-time fuel requirement of PHWRs having an aggregate capacity of 10,000 MWe which are planned to be installed in the next 15 years. Plutonium and depleted uranium recovered from the spent fuel of these PHWRs would be the starting point for launching the plutonium fuel cycle involving fast breeder reactors. A typical plutonium fuel cycle is shown in Fig. 11.13. It starts with the reprocessing plant which has already been discussed earlier and is followed by plutonium fuel fabrication. Pilot plant facilities for plutonium fuel fabrication have been set up at BARC and after developing the necessary technology, the fuel for India's Fast Breeder Test Reactor (FBTR) has been fabricated. A larger plant for meeting the plutonium fuel fabrication requirements for Prototype Fast Breeder Reactor is under construction at Tarapur.

Thorium Fuel Cycle

India has the world's largest resources of thorium and harnessing of those resources would ensure a steady supply of abundant nuclear power for many centuries. The third stage of India's nuclear power programme would be based on the use of thorium as fertile material and uranium-233/plutonium as fissile material. Even though the technology for thorium fuel cycle would be required only after a few decades an early start has been made to develop the necessary know-how for this purpose. India's first large research reactor CIRUS had thorium rods in the blanket region for the conversion of thorium-232 into uranium-233. The technology for reprocessing of the irradiated thorium rods from CIRUS was developed at Trombay and uranium-233 recovered has been used for setting up a small reactor PURNIMA-II to gain experience in the physics of the system. The technology for the fabrication of thoria blanket was established by fabrication of thoria blanket elements for FBTR. Efforts are also in progress to scale up the reprocessing of thorium based fuels at Kalpakkam. A new reactor KAMINI based on uranium-233 as fuel is proposed to be set up at Kalpakkam. These efforts would ensure that India would have a fully developed thorium technology by the time it is needed by the country.

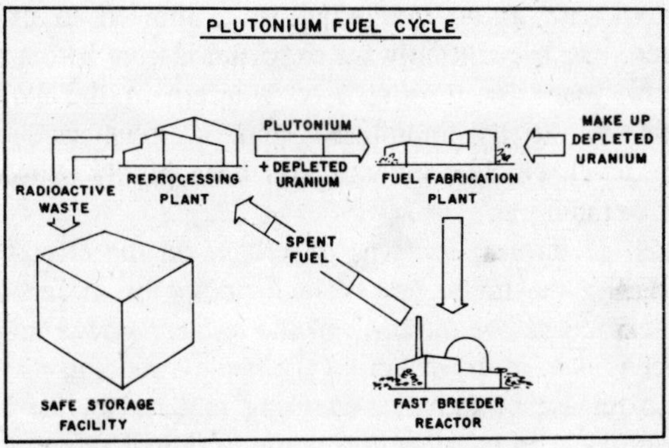

Fig. 11.13: Plutonium Fule Cycle.

Environmental Aspects

Before I close, I would like to mention briefly the environmental aspects of nuclear energy. It is true that radioactivity is dangerous, but we always have radiation hitting us from external and internal sources. Cosmic rays and radioactive materials which occur naturally in the earth's crust expose each one of us to about 100 millirem per year. Exposure to people staying near beaches which have monazite sand, such as those in Kerala, is of the order of 400 millirem per year. Naturally occurring radioactive elements which we take into our bodies through food, water, etc., lead to about 100 millirem per year of internal exposure. Further, diagnostic X-rays, television, etc., expose us to about 50 millirem per year. Exposure to this radiation has so far not contributed to any problems for us. Health surveys conducted by BARC in Kerala have not shown any adverse effects on people living in zones of high radiation background. Now let us compare this natural exposure with the level of artificial exposure. Nuclear weapons testing in the atmosphere contributed to about 5 millirem per year but this has now decreased to less than 0.5 millirem per year. Exposure to people living on the fence of Rajasthan Atomic Power Station is of the order of 2 millirem per year. You can clearly see that this is very small as compared to the natural background. But you may ask "what is the radiation hazard to the employees in nuclear energy installations"? This aspect has been discussed at the international level and the permissible levels of radiation have been defined by the International Council on Radiation Protection (ICRP). According to this, the radiation exposure to an individual isn't allowed to exceed 5,000 millirem per year, but normally the exposure is kept within about 1,000 millirem per year. The exposure to workers in the atomic energy installations is strictly monitored and is kept well below the permissible level. Nuclear industry has been highly conscious of the danger due to radiation.

A wealth of information was generated on the effects of nuclear radiation during the tragic use of nuclear bombs in Japan. As these events took place before the birth of the nuclear power industry, great care has been taken in the design of the nuclear reactors so that there is virtually no release of radiation-emitting material in the nuclear fuel cycle which has been instrumental in the excellent safety record of the industry during the last three decades. In spite of this, many people propagate that a nuclear reactor is dangerous and can explode like a bomb. In fact, a nuclear power reactor cannot explode. It is even impossible to make it explode like a bomb. All that can happen in the worst conditions is that a few fuel elements can melt which would be retained inside the pressure vessel or the floor of the containment building if pressure vessel also develops a leak. Because of this, even the worst accident of nuclear industry which occurred at Harrisburg did not result in a significant release of radioactivity to the environment. Not a single person received radiation in excess of that stipulated by ICRP and nobody died in that accident. Many scenarios have been built up by opponents of nuclear power to dramatize the dangers of nuclear power reactors and one of them is called the China Syndrome in which the entire core melts down, goes through the concrete and a significant fraction of radioactivity is released to the surroundings. These are however highly hypothetical scenarios and not actual possibilities. Because of these facts nuclear power has been accepted by most of the advanced countries. Today France has more than 50 per cent of its electricity coming from nuclear power stations. In many countries in Europe, like Finland, Belgium, the U.K., it is around 18 to 20 per cent. In Japan and West Germany more than 20 per cent power comes from nuclear stations.

Conclusions

Nuclear power generation is a high technology area and it requires a special culture. Nuclear power reactors have to be carefully designed, manufactured to stringent specifications, properly located and carefully operated throughout their lives. Our efforts have been to maintain a high level of competence by proper manpower training, proper selection of sites and following rules and regulations which are required to make its operation safe. The present tendency regarding siting is to co-locate many units in one area, e.g. Japan has four units of 1,000 MWe each located within one square kilometre of sea coast. Electricity can readily be transported efficiently by high voltage transmission lines and therefore location of reactors need not be tied up to the load centres. I am not an expert on nuclear sites. But as far as the country's nuclear power programme is concerned, you will see that it is not only our ability to build the first few power stations and operate them successfully, but also our capability in the technologies involved and the management of entire gamut of auxiliary industries from uranium production, fuel fabrication, fuel reprocessing to waste management that gives us confidence in launching a bigger programme. This is a technology in which the country is now completely self-sufficient which is not the case with many other areas even in conventional technologies. On the one hand we have this self-reliance in nuclear technology and on the other we hear that we may get experts from Sweden for afforestation. This dependence on foreign expertise must change. Well, that in short is the promise of nuclear industry for the country. Our uranium resources are sufficient at present and certainly more uranium will be found. We have plenty of thorium and therefore if we take a long-range view, nuclear power is the answer for meeting our energy needs. It has no limitations on resources like those you have for hydel power or the availability and transport of coal for thermal power. However, the problems associated with financial resources do exist. Regarding the sophistication of the technology, our country has many other areas of high technology such as air-transport systems, communication systems, etc., and therefore we should be able to run our nuclear power stations and meet all the power requirements of our country.

TWELVE

NUCLEAR POWER

M.R. SRINIVASAN

INTRODUCTION

Energy is a basic input to the national economy, both for agriculture and industry. Apart from being an instrument for improving the quality of life the electrical energy being the most convenient form for utilization, major importance is given to the production, transmission and utilization of this energy. The primary commercial energy inputs to economy of any nation are from coal, oil, water and progressively nuclear energy. The available renewable source of energy for bulk electricity generation is hydropower and this source must be exploited to the maximum possible extent. Unfortunately, hydro plants can be constructed only at specific locations and involve high capital costs and long gestation periods, due to their being multi-purpose schemes. The total annual estimated energy available in India from hydro-resources is 472 Twhrs and the utilization of hydro-potential today is only around 20 per cent. This is partly due to inaccessibility of locations, ecological problems and problems concerning sharing of river waters. Further as coal and oil have alternative uses, fossil fuels should be conserved for other uses, as far as possible. However, presently the contribution of coal to electricity production is substantial and this is likely to remain so during the next two decades. It is expected that over a period of time the contribution of nuclear energy is expected to become significant.

A number of countries have launched sizable nuclear power programmes. The USA, USSR, France, Canada, etc., are in the forefront with regard to power production using nuclear energy. The USA has the largest number of nuclear reactors presently in operation. A number of 440 MWe and 1,000 MWe capacity plants are in operation and under construction in the USSR. France has gone in a big way in nuclear power production utilizing 900 MWe and 1,300 MWe multiple units and its present share of nuclear electricity is 60 per cent of total generation. This is expected to go up to 75 per cent in the nineties. Japan has also started using nuclear energy for power production in a big way. Reactors of boiling water (BWR) type and pressurized water (PWR) type using enriched uranium as fuel are in operation in most of these countries. However, Canada has used the concept of Pressurized Heavy Water Reactors (PHWR) using natural uranium as fuel in sizes of 500 MWe, 600 MWe and 750 MWe. The Canadian reactors are performing very well. India has also adopted the PHWR type reactors for power production in the first phase.

At present the contribution of nuclear power to the total electrical power in the world is around 12 per cent. As many as 317 nuclear reactors of various capacities are presently operating in the world producing about 191,000 MWe of electricity. The production of nuclear electricity is expected to go up to about 257,000 MWe by 1987 with the commissioning of additional 187 units. Table 12.1 lists out the operating capacities of nuclear power plants in the world.

TABLE 12.1: **Operating Capacities of Nuclear Power Plants in the World**

*(Status 1983)**

Country	Operating Capacity MWe	
	1983	1987
United States of America	62400	122800
France	23400	52300
USSR	17900	40300
Japan	16600	26900
Federal Republic of Germany	9800	19200
Sweden	7300	9400
Canada	6700	11800
United Kingdom	6500	11600
Belgium	3500	5500
Taiwan	3100	4900
Spain	2000	13200
Switzerland	1900	2900
Bulgaria	1600	3600
Italy	1200	3200
Republic of Korea	1200	6500
India	1100	1700
Czechoslovakia	750	4100
Brazil	600	1800
Hungary	400	1600

* Nuclear Power Reactors in the *World Reference Data* Series No.2, International Atomic Energy Agency, Vienna, September 1983.

Indian Nuclear Power Programme

Now I propose to discuss the Indian nuclear power programme. We decided in the early sixties to build a number of heavy water natural uranium reactors similar to the Canadian ones, although our first station at Tarapur is of the boiling water type using enriched uranium. When the two units of 210 MWe at Tarapur were commissioned, they were the largest nuclear units anywhere in Asia (other than the Soviet Union). India is one of the few countries in the world and the only one among the developing countries to have its own technology for design, installation and operation of nuclear power stations as well as the associated fuel cycle facilities for heavy water production, uranium and zircaloy production, fuel fabrication, fuel reprocessing and waste management. We have chosen the pressurized heavy water reactor system to be the backbone of our nuclear power programme since (1) no uranium enrichment is required, (2) design is amenable to indigenization, (3) uranium requirement per MWe is less, (4) plutonium production is high and (5) there are built in safety features in the reactor design. Based on these considerations, two 220 MWe heavy water reactors were put up at the second nuclear power station at Rajasthan. The first Rajasthan reactor which is in operation for the past 12 years can be considered as a demonstration reactor as this is the first reactor of this type built in our country. Unit two of Rajasthan station has been performing quite well due to various improvements carried out based on lessons learnt from Unit-1 operation. The commissioning of the first unit at Kalpakkam, the third nuclear power station, near Madras during July 1983 marks the coming of age of the Indian Atomic Energy programme as the responsibility for engineering, construction, commissioning and operating the power station rested with Indian engineers and scientists. India became the ninth country in the world to develop the capability to execute nuclear power plants totally on its own. The performance of this indigenously built unit has been much better than the first unit at Rajasthan which was largely imported. This is evident from the performance of the unit during the first year of commercial operation when 1264 million units have been generated at a capacity factor of 61.3 per cent. The second unit at Kalpakkam is scheduled for commissioning in August/September 1985. The fourth atomic power station situated at Narora is in an advanced stage of construction. As the site is located in a seismic zone, extensive seismic analysis had to be carried out, for the buildings as well as equipment. In addition, several new design features were incorporated with the idea that some of them could be used for the 500 MWe reactor. Concreting works for two 235 MWe heavy water reactors at Kakrapar, the fifth nuclear power station, commenced in December 1984 and orders for most of the major equipment have been placed. The Government has approved construction of two units of 235 MWe units at Kaiga, a site in Karnataka, and addition of two more 235 MWe units in Rajasthan.

15-Year Nuclear Power Programme

The present total installed capacity is about 1,100 MWe and a further 1,200 MWe is under construction. Another 940 MWe has been approved for commencement.

A substantial increase in the nuclear capacity is planned in the immediate future. With the many achievements in engineering and fabrication of critical nuclear components to our credit, together with the extensive industrial and research infrastructure that has been developed, a stage has come for take off in the nuclear power generation and a target of 10,000 MWe of nuclear power by the turn of the century has been set. To achieve this target, it is proposed to install 12 units of 235 MWe in addition to those already under construction. Concurrently, design of a 500 MWe unit has also been taken up and 10 such standardized units are proposed for completion by the end of the century. Success of this programme requires much greater involvement of Indian industry which has to produce high quality equipment not only in time to meet project schedules but also economically. The total estimated capital investment of about Rs.14,000 crores needed for 10,000 MWe programme (at 1983 prices), though large, is only a fraction of the investment needed for the entire power programme for attaining the target of nearly 100,000 MWe to 120,000 MWe planned by the end of the century. The cost of nuclear power today compares favourably with coal-fired stations located 800 kms away from pithead and in the 1990s would be cheaper than coal-fired stations even at pithead.

Fast Reactor Programme

With the establishment of 10,000 MWe nuclear capacity based on natural uranium reactors, adequate plutonium will be produced annually making it possible to add 1,000 MWe of fast breeder reactors every year. This plutonium, together with Uranium-238 left over from the 10,000 MWe programme will enable putting up 350,000 MWe of fast breeder reactors in the next century. Thus, the installed nuclear capacity that can be supported by our somewhat limited uranium reserves can be increased almost 35 times. This is the target set for our fast reactor programme. A 15 MWe Fast Breeder Reactor (FBTR) is nearing completion at Reactor Research Centre at Kalpakkam. The design, construction and R&D experience generated at Kalpakkam Centre has provided the confidence to undertake the design and construction of a larger prototype fast breeder power reactor (PFBR) of 500 MWe capacity.

Safety in Nuclear Power Plants

Compared to chemical plants, the safety record of nuclear power plants is of a very high order. All are aware of the recent Bhopal tragedy and some of you may have read about the accident which took place at one of the reactors at Three Mile Island in the USA some years ago. While in the former case many people lost their lives, there was no casualty due to the nuclear accident at Three Mile Island. This is attributable to the importance attached to safety right from siting, design, commissioning and operation of nuclear power plants.

I would now like to touch upon the safety aspects of nuclear power plants. A conceptual design of the nuclear power plant including siting, hazard analysis of postulated accidents, measures proposed for tackling emergency situations, etc., is reviewed by an independent agency before undertaking detailed work. The design of a nuclear power plant provides for redundancy in plant cooling systems and ventilation systems. The control system of a nuclear reactor includes triplicated regulating and protective systems for regulation and safe shut-down as required.

Extensive and rigid quality assurance measures are adopted during design, material selection, fabrication of equipment, construction, commissioning and operation. In addition during the operational stage of the plant regular and periodic inservice inspection of equipment is carried out to detect in time any degradation in performance of equipment.

Strict control measures are adopted for minimizing radioactivity releases from the plant through water, air and land routes.

In the present standard design, the reactor is provided with double containment to ensure negligible release of radioactivity to the environment even during accident conditions. In order to mitigate the consequences of an unlikely accident, to the members of the public, the philosophy adopted is to provide 1.6 km exclusion radius around the plant fully. In this exclusion zone people are not allowed to habitat or carry on cultivation. We also provide a sterilized zone of 5 km radius beyond the exclusion zone where normal agricultural activities are allowed but growth of population and commercial activities are controlled.

A great deal of importance is also given to selection, recruitment, training/retraining and qualification of the operating personnel. This goes a long way in improving reliability and safety.

Waste Management

One of the important areas to be taken care of in production of power utilizing nuclear energy is handling and safe disposal of radio-active wastes generated. Operating power plants produce wastes in small quantities. These are mainly in the form of ion exchange resins which can be contained and disposed of. Some of the liquid wastes are concentrated by solar evaporation and the concentrated waste is fixed in concrete for burial. Also liquid waste is treated, diluted and is disposed off. In the fuel reprogressing plants some of the waste generated is more radio-active than that produced in nuclear power plants. These wastes are vitrified into glass mass and burried underground. Since the waste is fixed in glass there are no chances of leaching of radioactivity to subsoil water.

Energy needs in J & K

The State of Jammu and Kashmir with an installed capacity of 204 MWe consisting of 181 MWe hydro and 23 MWe thermal is a deficit State as far as power is concerned even after import of power. The situation is expected to improve after commissioning of Salal hydro-electric project. With the planned developmental activities the need for power in the State is likely to increase with time. The energy needs of J&K may have to be augmented by tapping additional hydro-sources including mini hydro-plants. Even though oil is expensive, in view of the geographical location, the immediate need can be met from thermal plants utilizing oil. Small capacity diesel generators may also have to be considered.

With the improvement of living standards there will be greater demand for low-grade heat for heating purposes. In developed countries dual-purpose plants for producing electrical power and low-grade heat utilizing fossil fuel and oil are well established. Medium and small-size nuclear reactor unit designs have already been developed and are being used in developed countries for electric power generation and district heating. Low temperature nuclear heating has been demonstrated in practice to be technically feasible and commercially viable. In fact pressurized water reactors (PWRs) using enriched uranium as fuel have been adopted for this purpose in many countries including the USSR. With a view to achieving the higher efficiency and economy it would be necessary to locate these nuclear units as close to population centres as possible. In developed countries designs have been evolved to enable siting of these plants close to metropolitan areas with minimum risk to the public. Design of such reactors for Indian conditions has yet to be taken up. The present standard design of nuclear plants used by us are meant only for electricity production as these have been engineered with power production as the ultimate goal. Considerable design effort and cost analysis are required to be carried out before dual-purpose reactors can be considered in our country.

To integrate even our 235 MWe PHWR reactor, which by international standards is a small reactor, the grid system requires to have a capacity of some 1,500 to 2,000 MWe. There should also be a minimum load at all times of the year of some 500 to 600 MWe to permit the nuclear unit to operate at base load. This condition is very important as otherwise the economics of operating the nuclear plant will be very adverse. It would appear that threshold conditions for economic integration of a nuclear power plant in the J & K State may not arrive for 10 or even 15 years.

Conclusion

Nuclear power will increasingly supplement coal and hydro power for generation of electricity in various parts of the country that are far away from coalmines. Ultimately these areas could possibly use nuclear energy as a main source for power generation. Efforts should be directed towards harnessing the available hydro-resources fully in all parts of the country.

In Jammu and Kashmir all the readily available hydro-potential should be tapped. This should be further augmented by medium-size high efficiency thermal plants utilizing oil and small-size diesel generators. In due course nuclear energy could play a role by using dual-purpose nuclear reactors for providing low-grade heat and electricity.

THIRTEEN

OPERATIONAL ASPECTS AND EXPERIENCE OF NUCLEAR POWER STATIONS

G.V. NADKARNY

INTRODUCTION

The progress of nuclear power technology has been phenomenal since the world's first nuclear power station started 30 years ago. In this brief review on the operational aspects and experience of nuclear power plants, I would like to highlight their global impact in general by indicating the total number of nuclear power stations in operation and their contribution to the total generation of electricity. This paper would also mention the main classifications or types of nuclear power stations and their special features that pose operational restraints as compared to the conventional thermal power stations.

Performance of nuclear power stations is judged by several criteria like operating factors, capacity factors, reliability in continuous operation (no. of outages), cost of power generated, safety of personnel and pollution effects on the environment.

In particular this presentation deals with the operational experience gained in the five operating units at the three operating nuclear power stations in India, i.e., at Tarapur Atomic Power Station (Tarapur), Rajasthan Atomic Power Station (near Kota) and Madras Atomic Power Station (Kalpakkam near Madras). It reviews the achievements, problems, the lessons learnt and the prospects based on our nearly 50 power reactor-year experience.

THE WORLD SCENE

Installed Capacity (Tables 13.1 and 13.2)

The world's first nuclear power plant, OBNINSK APS, located at Kaluga, USSR, and rated at 5 MWe went into operation in 1955 and ushered in the beginning of the era of nuclear power. By the end 1982, nearly 300 nuclear power plants in 30 countries were operating with a total capacity of 173,000 MWe providing over 9 per cent of world's electricity. Another 216 nuclear power plants were already under construction. In some countries contribution from nuclear power generation reached 16 to 40 per cent of total electrical generation. Typical present day standard size ratings of nuclear power plants are 220 MWe, 235 MWe, 440 MWe, 500 MWe, 650 MWe, 960 MWe, 1100 MWe and 1350 MWe.

Estimates (1982) for 1985 indicated that there would be about 427 nuclear power plants with a total installed capacity of 290,000 MWe. These would increase to about 508 power plants and 370,000 MWe by 1990 contributing 12 per cent of the total electricity generating capacity and to some 570 nuclear power plants and 400,000 MWe by the year 2000 A.D. contributing about 15 per cent of the total generating capacity. (in Figs. 13.1 & 13.2)

Classification of Nuclear Power Plants (Table 13.3)

Depending on the fuel, moderator and coolant used for the fission reaction in the nuclear reactor of a nuclear power plant, these plants have been classified into the following main types:

(1)	PWR	–	Pressurised Water Reactors.
(2)	BWR	–	Boiling Water Reactors.
(3)	GCR (MAGNOX)	–	Gas Cooled Reactors.
(4)	PHWR	–	Pressurised Heavy Water Reactors
(5)	FBR	–	Liquid Metal Cooled Fast Breeder Reactors.

Of the 297 nuclear power plants in operation at the end of 1982, typewise distribution was as is shown in Table 13.4.

TABLE 13.1: **Global Listing of Nuclear Power Plants (1982) and Capacity**

Nuclear power reactors in operation and under construction at the end of 1982

Country	In Operation*		Under construction*		Nuclear generated electricity in 1982 (estimated)+	
	Number of units	Total MWe	Number of units	Total MWe	TWh	%
Argentina	1	335	2	1292	1.9	7.0
Belgium	6	3473	2	2102	14.5	30.2
Brazil	1	626	2	2490	not available	
Bulgaria	4	1632	2	1906	9.0	25.0
Canada	14	7278	9	6310	36.4	9.7
Cuba	—	—	1	408	—	—
Czechoslovakia	2	762	6	2520	7.0	8.0
Finland	4	2156	—	—	15.9	40.3
France	32	23355	27	30200	103.0	38.7
German DR	5	1694	—	—	12.0	11.9
Germany FR	15	9831	12	13155	60.3	17.4
Hungary	1	408	3	1244	not available	
India	4	809	4	880	2.2	2.0
Italy	3	1232	3	1999	6.6	3.8
Japan	25	16587	11	10289	103.7	20.3
Korea, Republic of	2	1193	7	6227	3.8	8.5
Mexico	—	—	2	1308	—	—
The Netherlands	2	501	—	—	3.6	6.8
Pakistan	1	125	—	—	0.1	0.2
Philippines	—	—	1	620	—	—
Romania	—	—	2	1320	—	—
South Africa	—	—	2	1842	—	—
Spain	4	1973	11	10156	14.0	12.0
Sweden	10	7330	2	2100	37.3	38.7
Switzerland	4	1940	1	942	14.3	28.2
Taiwan	4	3110	2	1814	13.1	35.0
United Kingdom	32	6462	10	6052	41.6	16.4
USA	80	62376	61	67228	283.0	12.6
USSR	40	17219	31	30486	83.0	6.5
Yugoslavia	1	632	—	—	not available	
Total	297	173039	216	204780	—	—

DDR has officially indicated to the IAEA that no reactors are under construction.

Sources: * IAEA Power Reactor Information System;
+ OECG and NEI estimates.

TABLE 13.2: **Global Projections (As of 1982) of Nuclear Power Plants for 1985 and 1990**

Capacity projections, by country, 1985 and 1990*

Country	1985		1990	
	Number of units	Capacity MWe	Number of units	Capacity MWe
Argentina	2	935	3	1,627
Belgium	8	5,485	8	5,485
Brazil	1	626	3	3,116
Bulgaria	5	2,585	6	3,538
Canada	20	11,076	24	14,469
Cuba	—	—	1	408
Czechoslovakia	8	3,282	8	3,282
Finland	4	2,156	4	2,156
France	51	44,195	60	54,845
Germany DR	5	1,694	5	1,694
Germany FR	21	16,432	27	22,986
Hungary	3	1,224	4	1,632
India	6	1,249	9	1,909
Italy	4	1,267	6	3,231
Japan	34	24,705	36	26,876
Korea, Republic of	7	5,534	9	7,420
Mexico	1	654	2	1,308
The Netherlands	2	501	2	501
Pakistan	1	125	1	125
Philippines	1	620	1	620
Poland	—	—	1	440
Romania	—	—	2	1,320
South Africa	2	1,842	2	1,842
Spain	9	6,460	15	12,129
Sweden	12	9,430	12	9,430
Switzerland	5	2,882	5	2,882
Taiwan	6	4,924	6	4,924
United Kingdom	38	10,062	42	12,514
USA	110	94,288	132	118,979
USSR	60	36,728	71	47,706
Yugoslavia	1	632	1	632
Total	427	291,593	508	370,025

Source: IAEA Power Reactor Information System.

* Only those reactors currently known to be connected to the grid or under construction are considered.

Performance of Nuclear Power Plants

The nuclear power plants have shown impressively high load factors (capacity factors) throughout their operating life. Cumulative load factors which are indicative of their overall performance are shown in Table 13.5 based on a total of 226 nuclear power plants with an installed generating capacity of 168,500 MWe.

The moving annual load factors of the top five rankers in September, 1984, varied from 94.8 per cent to 99.1 per cent and three of these were the Bruce-1, Bruce-2, Bruce-3 units of PHWR type.

The cost of power generated from nuclear power plants has been competitive compared to cost of power generated from coal and oil in many countries especially for nuclear power plants of 500 MWe rating and above.

The effect of nuclear power plants on the environment by way of pollution, through discharge of radioactive effluents, has been minimal due to the stringent design and operational features incorporated in the plants. Similarly over-exposures to radiation workers and fatalities due to nuclear power plant accidents are almost nil compared to the total number of radiation workers involved in these plants and more than 3,000 nuclear reactor-years experience of nuclear power operation gained so far.

SALIENT FEATURES OF NUCLEAR POWER PLANTS

Precautions against Radioactive Core

The reactor core as a result of fission reaction during the plant operation accumulates large amounts of highly radioactive fission products within the fuel elements. These need to be kept cooled all the time, even during shutdown, to prevent damage to the integrity of fuel-cladding and consequent escape of radioactivity. This makes it mandatory for the control room to be manned by qualified operators throughout the life of the reactor. A nuclear power station therefore cannot be shut down, locked and left to itself unattended during weekends, vacations or due to any reasons like shortages of operating supplies or even strikes unlike in other industrial plants. The reactor power has to be always monitored with proper instrumentation, provided with protective devices, even during shutdowns.

This same consideration of reactor safety at all times also makes it necessary to ensure the availability of the following, backed up with sufficient redundancy:

TABLE 13.3: **Classification of Nuclear Power Plants**

TYPE	LWR		PHWR	MAGNOX (GCR)	LMFBR
	PWR	BWR			
Fuel	Enriched Uranium 1-3%	Enriched Uranium 2-3%	Natural Uranium	Natural Uranium	Plutonium + Enriched Uranium
Cladding	Zircalloy	Zircalloy	Zircalloy	Al.-Mag	Stainless Steel
Coolant	Light water (indirect)	Light water (Direct)	Heavy water	Carbon dioxide	Sodium
Moderator	Light water	Light water	Heavy water	Graphite	Nil

LWR = Light Water Reactor
PWR = Pressurized Water Reactor
BWR = Boiling Water Reactor
PHWR = Pressurized Heavy Water Reactor
MAGMX = Magnesium Aluminium-clad Natural Uranium Fuelled $eactor.
LMFBR = Liquid Metal-cooled Fast Breeder Reactor.

TABLE 13.4: **Typewise Per cent Distribution of Nuclear Power Plants (December 1982)**

Type	Per cent of Total Nuclear Puwer Plants
PWR	56.5
BWR	25.6 (TAPS Type)
GCR (MAGNOX)	11.9
PHWR	5.0 (RAPS, MAPS Type)
FBR	1.0

TABLE 13.5: **Average Cumulative Load Factors (September 1984)**

Type	No. of Plants	Capacity (MWe)	Arithmetic Average Percentage	Capacity Weighted Average Percentage
PHWR	18	10,000	65.0	70.0
GCR (MAGNOX)	26	8,500	60.0	57.8
PWR	122	103,000	59.9	58.8
BWR	60	47,000	58.2	57.7
	226	168,500		

(i) Emergency Power Supplies (Diesel Generating Sets, Motor Generating Sets, Battery Banks),
(ii) Heat Transport Shutdown Cooling Pumps, Moderator Cooling Pumps,
(iii) Emergency Cooling Pumps,
(iv) Three independent sources of grid power supply to ensure uninterrupted power,
(v) Routine testing of process and regulating systems, protective systems and safety systems,
(vi) Monitoring and control of effluents of radioactive discharges in air and water routes,
(vii) Emergency safety systems like emergency injection and containment,
(viii) Integrity of physical barriers to prevent escape of radioactivity to public domain,
(ix) Administrative controls to ensure radiation hazards control procedures, emergency drills and sector emergency evacuation procedures,
(x) Maintenance of minimum water level in cooling water supply source such as lake (RAPS) and hence minimum submergence level in the pump house, and
(xi) Adherence to Technical Specifications and other instructions issued by the Atomic Energy Regulatory Body.

Work Procedures

Proper work methods and procedures to prevent bypassing of safety during maintenance or operation.

Radiation Exposure Control

Monitoring and control of Radiation Exposures to station staff and other radiation workers by proper planning of work in advance in mock-ups and drills and use of protective clothing. Investigation of over-exposures and remedial action to prevent recurrence.

Base Load Operation

Because of high capital cost, long gestation period for licensing, construction and commissioning, the nuclear power plants have to run as base load stations. This is also necessary to prevent fuel failure and release of fission products into the coolant which would otherwise result from frequent load changes to suit the grid load demand.

Low Fuelling cost of Nuclear Fuel and Refuelling of Reactors and Handling of Highly Radioactive Spent Fuel

Nuclear fuel is a very compact energy-loaded fuel. One kg. of uranium gives the heat equivalent of 25,000 kg. of coal in today's nuclear power plants even though only one per cent of that one kg. of uranium is used up in the fission reaction. Low fuelling cost of nuclear power plants off-sets the high capital cost in the long run and is a major plus factor in their economic viability.

In light water reactors and gas cooled reactors the refuelling is done during shutdown while in PHWR plants on power refuelling is done. The discharged fuel has to be handled remotely with shielding, due to its highly radioactive nature. Once fuelled, nuclear power plants can, then on, operate for one to one-and-half years (PWR and BWR) while PHWR's can operate one to two months on available reactivity which is limited and is made up by on-power fuelling.

Fuelling is done as per computer programmes to ensure uniform flux flattening and prevent hot spots in the reactor core.

Heavy Water Escapes, Recoveries and Losses

In PHWR, the success of the technology depends on keeping the escapes and losses low while making most recoveries of the heavy water spills and escapes. The downgraded (0.3 to 90%) recoveries have to be upgraded to 99.7 per cent and above by distillation columns. In developing countries with D20 shortage, this is very important. Heavy water costs have been steadily going up and presently are in the range of Rs.3000-3500/kg. A nuclear power plant needs about 250 tonnes of reactor grade heavy water and some of it (60 to 70 tonnes) is at 87 kg/cm^2 pressure and 260 to 298°C temperature.

Tritium and Argon Hazards

Part of the heavy water recirculated in the moderator system as well as the primary coolant system gets converted to tritium, a radioactive isotope of deuterium. The atmosphere in the reactor building is thus usually contaminated with tritium and this poses a health hazard to operating personnel. Similarly the air exposed to radiation in the vicinity of the reactor core gets radioactive with Argon-41 radioactivity whose hazards on pesonnel can only be reduced by proper control of ventilation air flows in active areas.

TABLE 13.6: **Annual Performance of five World Rankers Among 253 Nuclear Power Plants**

(As of September 1984)

Annual Ranking	1	2	3	4	5
Unit Name	BRUCE-3	BRUCE-4	OCONEE-1	TRI-CASTIN-1	BRUCE-1
Country	Canada	Canada	USA	France	Canada
Type	PHWR	PHWR	PWR	PWR	PHWR
Moving Annual Load Factor %	99.1	99.0	96.2	95.3	94.8
Cumulative Ranking	1	2	106	30	5
Cumulative Load factor	88.4	87.3	61.4	75.0	82.4
MWe Gross	826	826	926	957	826

Inaccessibility of Equipment

Major and critical equipment like main heat transport pumps, boiler feed water control valves, reactivity mechanisms, heavy water vapour recovery units and quite a number of safety instruments are located in boiler room, moderator room, etc., which are called *Shutdown* areas. In this area which forms 80 per cent of the space in reactor building, access to equipment is prohibited during unit operation due to existence of high radiation fields. It therefore calls for a high degree of maintenance and trouble-shooting during shutdown periods.

All operation and monitoring of equipment including on-power fuelling machines in shutdown area is done by remote sensing and control.

Poison Over-ride and Poisoning Out of Reactor and Need for Uninterrupted Grid Power Supply

In PHWR type of reactors, on-power fuelling provides the advantage of avoiding long refuelling outages so characteristic of PWRs and BWRs. However this puts a limitation on the total excess reactivity available for overriding the poison effect of fission products like xenon in PHWRs. As a result of this, all PHWR reactors, following a trip at full power, have to be restarted within the poison over-ride time of 35 minutes provided by the absorber rods. If this is not achieved, one has to wait 36 hours thereafter, for the xenon to decay, before the reactor can be started up. This is not the case with other types of nuclear power plants, i.e. BWR (TAPS). Every poison outage in PHWR results in unavailability of the unit for two to three days.

Chances of start-up of PHWR within the poison over-ride time of 35 minutes are increased if grid supply remains uninterrupted during unit outage.

INDIAN SCENE

Choice of Reactor Type

Of the five types enumerated, India has selected PHWR technology for its first stage of the long-term nuclear programme for the following reasons:

(a) Fuel Efficiency.
(b) Viability of indigenous manufacture of reactor components.
(c) Use of indigenous fuel and fuel manufacture.
(d) Use of indigenously produced heavy water.

There has been progressive indigenization in PHWR technology starting from RAPS-1 and proceeding through RAPS-2, MAPS-1, MAPP-2, NAPP-1, NAPP-2, KAPP-1 and KAPP-2 and so on.

Tarapur Plant (TAPS) was chosen from BWR technology as, at that time, PHWR was not fully developed and gas-cooled reactors were most expensive. The need then was to start Indian engineers and operators in gaining experience in building and operating nuclear power plants and check functioning of nuclear power plant in the Indian grid.

Operating Nuclear Power Plants

As of June 1985, five nuclear power plants are operational in India (in Table 13.7).

Contribution as of December 1984 from nuclear power generation to the total installed capacity in India is only 2.7 per cent (1,095 MWe in a total of 40,280 MWe of which thermal is 24,966 MWe and hydel is 14,219 MWe).The sixth nuclear power plant MAPS-2 (235 MWe) is about to be commissioned soon.

TABLE 13.7: **Indian Operating Nuclear Power Plants (June 1985)**

Nuclear Power Plant	Type	Capacity	First Critical	First Synchroni-zation	Declared Commercial
TAPS-1	BWR	210 MWe	1.2.69	1.4.69	9.10.69
TAPS-2	BWR	210 MWe	27.2.69	1.4.69	9.10.69
RAPS-1	PHWR	220 MWe	11.8.72	30.11.72	16.12.73
RAPS-2	PHWR	220 MWe	8.10.80	1.11.80	1.4.81
MAPS-1	PHWR	235 MWe	2.7.83	23.7.83	27.1.84

Nuclear Power Plants Under Construction and Planned

At Narora (U.P.) NAPP-1 (235 MWe), NAPP-2 (235 MWe) are in an advanced stage of construction and are expected to be operational by 1987 and 1988. Similarly at Kakrapar (Gujarat) KAPP-1 (235MWe) and KAPP-2 (235 MWe) construction work has started in 1981 and is expected to be completed by 1990 and 1991.

It has been planned to install 12 × 235 MWe units and 10 × 500 MWe units to achieve a total installed nuclear capacity of 10,000 MWe by 2000 A.D. As part of this plan, 2 × 235 MWe units as RAPS-3 and RAPS-4 and two similar units at Kaiga (Karnataka) were announced for construction in March 1985.

Brief Summary of Performance of TAPS, RAPS and MAPS

Tarapur Atomic Power Station (TAPS)

The main object of installing TAPS-1 and TAPS-2, i.e. giving Indian engineers and operators early experience of operating nuclear power plant in the Indian grid conditions has been fully met with creditable success. TAPS has completed 15 years of successful operation and accumulated 30 reactor years of power reactor experience.

Its average cumulative availability (operating factor) for the 16-year period till end-May 1985 has been 70 per cent and average cumulative capacity factors around 50 per cent. It has done better than most BWR nuclear power plants of its age. It has proved economically viable and has created operating surplus in most of the years. It has been one of the most reliable power stations supplying power to Gujarat and Maharashtra. TAPS-2 operated almost continuously from April 29, 1984, to end-December after it was restarted following the refuelling outage.

Rajasthan Atomic Power Station (RAPS)

RAPS-1

(a) RAPS Unit-1 which was the first PHWR plant in India and the forerunner of the type committed for the Indian nuclear power programme gave good experience to Indian engineers and operators to construct, commission and operate the Unit. It has completed almost 13 and half years of operation, as of end-May 1985.

(b) Its average cumulative availability for seven-and-three-quarter-year period from January 1974 to September 1981 has been 57 per cent and cumulative capacity factor 38 per cent. From September 20, 1981, since the south end shield leak started it remained shut down for four months in 1981 and except for brief 36 days in 1982, for the whole of 1983, 1984 and January 1985. It was resynchronized on February 1, 1985, after 35 months.

(c) Since its start-up RAPS-1 has performed well for the months of February, March, April and May 1985 with availability and capacity factors as shown in Table 13.8.

Its longest run in the four months was of 53 days since its recent start-up.

TABLE 13.8: **RAPS-1 Performance Since Start-up of February 1985**

Months (1985)	Operating Factor (%)	Capacity Factor (%)
February	33	13
March	69	38
April	100	72
May	62	40

RAPS-2

(a) RAPS-2 performance has been quite satisfactory. It has completed four-and-half-years of total operation as of end-May 1985.

(b) Its average cumulative operating factor is 66.3 per cent and average cumulative capacity factor is 43.3 per cent. The main reason for low capacity factors is the de-rating of the unit to 100 MWe due to safety reasons (plugging of 50 per cent of tubes in shutdown cooling heat exchangers) and later due to delay in refuelling after reaching equilibrium conditions.

Madras Atomic Power Station (MAPS)

MAPS-1

MAPS-1 is the first Indian PHWR which is totally an Indian product from design through construction, commissioning to operation. Its performance has been excellent. In the very first year of its commercial operation (1984) it achieved an annual operating factor of 72 per cent and capacity factor of 61.3 per cent which are even better at 75.3 per cent and 64.3 per cent respectively from the date of commercial operation (27.1.84).

Tarapur Atomic Power Station (TAPS) Experience

General Description

Located at Tarapur 100 kms north of Bombay, TAPS was the first nuclear power station in Asia and the first twin unit BWR station in the world. It has two 660/770 MWt (200/210 MWe) dual cycle BWR type reactors of very early design similar to Dresden-1 which has recently been permanently shutdown. Though the twin reactors having been housed in one building have resulted in overall cost reduction, this has imposed certain restrictions in operational and maintenance activities. Similarly the dual cycle design has also caused restraints in power raising during start-ups after refuelling control rod sequence changes.

The coolant is light water and hence cheap but fuel is 2-3 per cent enriched uranium which has to be imported and has caused problems due to restrictive trade policy of the USA, the original supplier. The steam is directly coming from the reactor to the turbine and hence the turbine needs shielding against fission products which may deposit on the turbine blades. The coolant gets full flow purification through ion exchangers.

The station was constructed as a turnkey contract by General Electric Co., USA, in the shortest time of four-and-half-years. The reactors have provision for large excess reactivity through off power refuelling to provide for long operation. This helps to over-ride xenon poison on reactor trip. It can be restarted like any thermal plant within hours unlike PHWRs. It is provided with 69 control rods and 284 fuel assemblies. The coolant system gets contaminated with fission products which result in radiation fields. Failed fuel (sheath rupture) cannot be removed on power or during short shutdowns unlike in PHWRs.

TABLE 13.9: **Operating Factors and Capacity Factors (1969-1984)**

Unit	Operating Factor (Per cent)		Capacity Factor (Per cent)	
	Range	Average	Range	Average
TAPS-1	57 - 93	80	35 - 63	49.3
TAPS-2	55 - 85	67	11 - 68	48.9

Operating Factors, Capacity Factors and Outages

The performance of both TAPS-1 and TAPS-2 has been consistently excellent for all the 15 years from 1969 to 1984. The range and average values are given in Table 13.9.

Detailed yearwise figures are shown in Table 13.10. The small number of outages and total absence of poison shutdowns have helped maintain good operating factors and capacity factors.

Outages

The number of outages per year is quite low as compared to RAPS and MAPS unit outages. In 1984, TAPS-1 had only six outages of which one was forced outage and five were planned. TAPS-2 had five outages four of which were planned and one was forced. TAPS-1 outage numbers varied from 4 to 22 per year and TAPS-2 outage numbers were between 2 and 24 per year for the period 1969 to 1984.

Refuelling and Annual Maintenance Outages

Annual major maintenance shutdowns are normally combined with refuelling outages. Till end-May 1985 there have been a total 18 refuelling outages, 9 in TAPS-1 and 9 in TAPS-2. The duration of refuelling outages has been progressively reduced to less than 100 days per unit. The strategy is to plan one refuelling outage in the station (either Unit-1 or 2 as applicable) and extend fuel cycle period of a unit to 18 to 20 months.

Fuel cycle performance in terms of period between successive refuelling outages has been exceedingly good as can be seen in Table 13.11.

Major Causes of Unavailability

Leaving aside refuelling and annual maintenance reasons, there have been major causes for unavailability of the units during the past 15 years in Table 13.12.

Major Equipment Problems

The following were the major equipment problems faced at TAPS:

Nuclear Reactor Systems

Reactor Vessel Internals (1971): Peripheral guide tubes were found displaced in TAPS-1 and these were repaired. Hold-down arrangements on all 20 such guide tubes were modified – 1971. In-core instrument bracings had broken in Unit-1. These were removed in TAPS-1 and 2 subsequently.

TABLE 13.10: **Year-wise Listing of TAPS Operating Factors, Capacity Factors and Energy Output (1969 - 1984)**

Year	TAPS-1				TAPS-2			
	Operating factor %	Capacity factor %	Energy output MW	Outages	Operating factor %	Capacity factor %	Energy output MW	Outages
1969 (Nov & Dec.)	93.3	53.6	165	6	85.6	49.8	153	5
1970	85.2	58.7	1079	13	78.5	59.7	1098	24
1971	59.0	46.1	843	5	67.8	51.5	947	5
1972	57.3	35.4	652	15	22.6	11.8	218	2
1973	58.9	41.2	757	18	84.5	67.9	1250	17
1974	79.2	45.2	832.	22	45.9	32.5	597	19
1975	66.5	50.4	927	3	64.9	50.3	926	7
1976	86.7	62.7	1157	12	74.8	61.7	1138	7
1977	70.0	54.1	995	17	85.1	68.2	1253	7
1978	73.4	51.5	941	11	76.2	62.3	1146	11
1979	81.6	52.5	966	6	71.0	54.0	993	5
1980	67.8	48.5	894	6	78.4	48.8	900	13
1981	68.4	43.2	794	13	77.1	52.4	964	9
1982	89.9	60.5	1112	14	55.4	30.3	557	10
1983	52.4	37.1	510	7	89.3	59.8	823	6
1984	87.8	49.3	909	6	64.1	47.0	868	5
Average	80.2	49.3	13757 (Total)	174 (Total)	67.3	48.9	14076 (Total)	152 (Total)

Note: Commercial Operation from 9.10.1969
Figures quoted Jan. to Sept. 1983 as Oct.-Dec., figures were not available at the time of this compilation at RAPS.

TABLE 13.11: **Fuel Cycle Performance**

Fuel Cycle No.	1	2	3	4	5	6	7	8	9
Cycle Days TAPS-1	654	275	609	642	426	459	464	438	587
Cycle Days TAPS-2	873	431	351	282	415	370	539	448	445

TABLE 13.12: **Major Causes of Unavailability at TAPS (1969-1984)**

TAPS-1	TAPS-2
(1) R.W. Problems (1971) (Rad Waste)	(1) Turbine bearing repairs (1971)
(2) Dry well leaks (1972)	(2) Dry well leaks (1972, 1973)
(3) S.S.G. repairs (1976)	(3) Turbine oil line repairs (1973, 1978)
(4) Condenser tube leaks (1975, 1981)	(4) S.S.G. repairs (1976)
(5) Recirculation pump seal (1977)	(5) C.W. system repairs and
(6) Recirc. pump cabling	(6) Condenser leaks (1978)
(7) Leaks in pipe tunnel (1980)	(7) Recirc. pump seal replacement (1982, 1983)
(8) Switch-yard bus fault (1982)	(8) Recirc. line leaks (1980)

Repairs to Dryer Separator Assembly Bottom Guide 1976: TAPS-2 seating was improper due to failure of weld of bottom guide bracket, weld repair done with Navy divers working underwater.

Secondary Steam Generator Leak Repair: There are a total of four SSG's. Location of leaky tube and plugging is difficult because of single manway access.

SSG 1A leaks repaired in 1973, 1975.
SSG 1B leaks repaired in 1976, 1978.
SSG 2A leaks repaired in 1974, 1976.
SSG 2B leaks repaired in 1974, 1976, 1978, 1979.
Radiation fields - 5 R to 30 R/hr.

Inservice Inspection: Stress corrosion cracking of S.S. components and piping during construction, SS cladding on reactor vessel and SSG channel, bottom control rod drive entry stubs, tubes area and nozzle safe ends. These were ground and all ends above 4" diam were overlayed with SS 308. These were inspected after five years during inservice inspection and found OK.

Small Pipe Failures: 2" Sch 80 pipe SS-304 piping in TAPS-1 and 1" drain pipe in clean up in TAPS-2. This was due to chloride pickup.

Large Pipe Failures: 8" size 80 emergency condenser piping in 1979 showed failure due to stress corrosion. Next major failure was in 6" SS-304 Sch 80 in TAPS-1 in 1980. Third failure was in TAPS-2 8"/10" SS-304.

Conventional Systems

Turbine

Turbine Bearing Failure (1971): Caused by failure of DC power supply to emergency bearing oil pumps in TAPS-2. Babbit was wiped out. Damaged bearings were replaced.

Turbine Control System Failures: Mechanical pressure regulators caused several forced outages. Modifications done to overcome problems.

Turbine Hydraulic Oil Failure and Stop Valve Support: Failures 1970-1973 at union joints and socket welds. Locking strips were welded to prevent rotation of union joints now checked in each outage. Socket weld failure of oil lines inside guard pipe failed due to vibrations. Extra clampings have been provided to overcome fault.

Turbine Causing Flange Leak due to Corrosion in TAPS-1: Repaired by welding and grinding. Moisture accumulation source is eliminated by increasing stage moisture draining.

Condenser

(a) Sea water used in (A1 brass/A1 bronze tube sheet) tubes. Erosion effects seen. Rectification done by (i) plugging in TAPS-1 and TAPS-2, (ii) Ferrous sulphate solution injection in CW inlet, (iii) sacrificial MS electrodes in water boxes, (iv) plastic inserts at tube inlet, (v) limiting water velocity, (vi) Regular inspection. Now performance is best considering sea water.

Failure of Main Transformer (1973)

Due to sea water ingress. Transformer was repaired with new coils at site. Corrective action taken to ensure oil pressure higher than sea water pressure, for all major transformers.

Switch Yard

This is open and exposed to Arabian sea. In the monsoon switchyard faces heavy saline pollution on insulators due to salt in moisture-laden sea winds. 20 trippings occurred in 1973 in 18 days due to flashover. Corrected by hot-line washing of insulators and replacing insulators with polycone insulators. Copper conductors used in place of ACSR conductors.

Significant Modifications

The following modifications were done:

(a) Replacement of Event Sequence Recorder by BARC-designed ESR.
(b) Augmentation of Rad. waste facility.
(c) Augmentation of condensate polishing capacity.
(d) Part augmentation of spent fuel storage bay since 1976. Now pending is 'AFR' (Away From Reactor) storage for spent fuel.

Impact of Grid

The enquiry specification of motors, transformers and other electrical equipment at TAPS included enough margins to enable the equipment to withstand greater variation in voltage and frequency of grid supply (47.5 cps and $\pm$ 10 per cent rated voltage). Auxiliary power design takes into account actual operating voltages expected at the equipment. Hence TAPS auxiliaries withstand grid fluctuations better than at RAPS.

Feedback from TAPS Operational Experience

(a) Enclosed switch-yard for MAPS.
(b) Bypass (Dump) condenser facility at MAPS.

(c) Vapour suppression pool for MAPS instead of dousing tank for quenching steam from hot pipe ruptures.

RAPS EXPERIENCE

General Description

Located 60 kms from Kota in Rajasthan, RAPS is the second nuclear power station in India but the first of PHWR type. It is situated on the banks of Rana Pratap Sagar (Chambal river). It has two 693 MW thermal (PHWR) reactors and the turbines are rated at 220 MWe each. The station was based on the design of the Canadian Proto-type PHWR nuclear power plant at Douglas Point, which went into operation in 1966. Construction on RAPS started much before Douglas Point was permanently shutdown last year (1984). Each reactor at RAPS is housed in a separate concrete Reactor Building with airlocks.

The coolant in heavy water is an expensive and rare commodity. The fuel is natural uranium dioxide which is Indian. Fuel has never been in any shortage at any time at RAPS. It is refuelling which has not been able to keep pace with operation. Steam boilers and heavy water circuits are located in the Reactor Building which is a containment building. This has to stand the stipulated leak tightness tests for clearance before start-up of unit due to safety requirements.

The Department of Atomic Energy, Government of India, took full responsibility for the construction of the station. Design was by Atomic Energy of Canada Ltd. Indian engineers fully participated in the construction, commissioning and operation of the two units. RAPS-1 has 60 per cent imported equipment and RAPS-2 has 60 per cent Indian equipment. RAPS-2 provided the first opportunity for Indian manufacturers to build major components like reactor vessel calandria, dump tank, boilers, fuelling machines.

Operating Factors, Capacity Factors and Outages

The performance of RAPS-1 has not been very satisfactory though for limited periods as in 1976, 1979 and 1980 it performed very well. RAPS-2 performance has been quite satisfactory for 1981, 1983 and 1984. Its capacity factors in 1981 in spite of the high operating factor of 80 per cent remained low at 42 per cent because of derating the unit to 120-150 MWe because of safety reasons. The range and average values are given in Table 13.13.

TABLE 13.13: **RAPS Operating and Capacity Factors**

Unit	Operating Factor (%)		Capacity Factor (%)	
	Range	Average	Range	Average
RAPS-1 (1974-84)	0.0 – 82.4	38.22	0 – 63.6	26.9
RAPS-2 (1981-84)	41.7 – 80.4	66.3	23 – 55.2	43.3

Detailed yearwise figures are shown in Table 13.14. The large number of outages and poison outs have contributed to low operating capacity factors.

Outages

RAPS-1

(a) The number of outages in RAPS-1 have totalled some 255 from the year 1973 to end of 1984. The longest outage lasted almost 1060 days (3 years) from 4.3.82 to 31.1.85 caused by the light water leak in south end shield. The next largest outage lasted 15 months, July 1977 to September 1978 caused by a labour strike during the annual maintenance shutdown.

(b) The number of outages have steadily decreased over the years from 70 to 13 per year. In its best year of performance 1979 and 1980 the number of outages were 14 and 19 respectively. From 1974 to 1984 almost 60 per cent of the outages were poison outages and 25 per cent of the outages were caused by the grid. There have been 181 grid power failures at RAPS-1 from 1973 to 1984.

(c) Subsequent to paralleling Chambal grid with Northern (BBMB) grid in January 1975, it was expected that the number of poison shutdowns would decrease. Though the grid sustained on most of the occasions of RAPS-trips, the unit got poisoned out on several occasions because the reactivity reserve was less than required due to the backlog in refuelling. From 1975 to 1984, the ratio of poison outages to total outages remained almost at 61 per cent.

RAPS-2

(a) The number of outages in RAPS-2 have totalled some 79 from 1980 to 1984. The duration of the largest forced outage was 147 days due to turbine blade failure from January 1982 to June 1982. This period was also used for annual shutdown work.
(b) Of the 79 outages 17 per cent were due to grid and 67 per cent were poison shutdowns. There were 33 grid power failures.

TABLE 13.14: **Year-wise Listing of RAPS Operating Factors, Capacity Factors and Energy Output (1973-1984)**

Year	RAPS-1			RAPS-2		
	% Operating factor	% Capacity factor	MW Energy output	% Operating factor	% Capacity factor	MW Energy output
1973	32.6	12.0	253			
1974	53.5	36.8	728			
1975	43.6	33.2	654			
1976	65.2	44.3	901			
1977	37.8	26.4	530			
1978	17.5	8.5	177			
1979	82.4	63.6	1252			
1980	72.2	54.2	1048	32	11.1	36
1981	42.6	26.3	503	*80.4	41.8	792
1982	5.7 (Long S/D)	2.6	50	41.7	23.0	433
1983	(Long S/D)	—	—	76.2	55.2	1023
1984	(Long S/D)	—	—	66.8	53.5	995
Average (1973-81)	49.7	34	6046	66.3 (1981-84)	43.4	3243

Commercial Operation Date	RAPS-1	RAPS-2
	16.12.1973	1.4.1981

*Best performance year

TABLE 13.15: **Raps Outages**

Year	RAPS-1			RAPS-2		
	Total outages	Poison shutdowns	Outages due to grid	Total outages	Poison shutdowns	Outages due to grid
1973	70	(Low power operation)	14			
1974	49	24	7			
1975	31	15	7			
1976	35	18	9			
1977	11	8	6			
1978	9	5	4			
1979	14	10	5			
1980	19	14	4	15	6	1
1981	13	10	4	24	18	4
1982	4	3	—	9	6	1
1983	Long S/D	—	—	15	12	4
1984	Long S/D	—	—	16	11	3
Total	255	107	60	79	53	13
In per cent (1973-82)	100 (255 outages)	42	24	100 (1980-84)	67	17
In per cent (1974-82)	100 (185 outages)	58	25	—	—	—
In per cent (1975-82)	100 (136 outages)	61	29	—	—	—

Annual Maintenance Shutdowns

The Annual Maintenance Shutdowns, which are supposed to be taken for four to six weeks every year for each unit, have not been regularly taken for two main reasons:

(i) Constant pressure from the Rajasthan State Electricity Board (RSEB) to keep RAPS units operating under the excuse of "power shortages in the State". However, keeping in view the larger grid (over 5,000 MWe thermal and 5,000 hydel) to which RAPS is connected, RAPS should learn to overcome such pressures from the grid and should ensure regular maintenance shutdowns.

(ii) Combining such shutdowns by RAPS with major equipment failures like turbine failures.

The annual shutdowns are shown in Table 13.16.

TABLE 13.16: **RAPS Annual Maintenance Shutdowns**

	RAPS-1 (1973-1984)		RAPS-2(1980-1984)	
1st *A.M.S.D.	July 1974	99 days (combined with turbine blade failure).	Jan. 1982	147 days combined with turbine blade failure).
2nd A.M.S.D.	July 1975	165 days (combined with turbine blade failure).	Sept. 1984	39 days (after 26 months)
3rd A.M.S.D.	July 1977	Long shutdown (Strike)		
4th A.M.S.D.	Sept. 1981	Long shutdown (end shield leak)		

A.M.S.D. = Annual Maintenance Shutdown.

Major Causes of Unavailability

Leaving aside Annual Maintenance Shutdowns, there have been other major causes for unavailability of the units in Table 13.17.

TABLE 13.17: **Major Causes of Unavailability of units**

RAPS-1 (1973-84)	RAPS-2 (1980-84)
(1) Turbine bearings (1973, 74, 75, 76) and blade failures.	(1) Turbine reheaters (1980, 1981, 1983).
(2) Channel Temp. monitoring instrumentation (1973)	(2) Thermal shield (1981, 1983)
(3) Deaerator (1973).	(3) Turbine blade failure (1982 & 83)
(4) Labour strike (1977-78)	(4) Generator (1984).
(5) Heat transport (1979, 1981 & 82)	
(6) Spurious H20 dousing (1981).	
(7) Moderator HXs (1980)	
(8) South end shield (1981-84).	

The systemwise allocation of unavailability is shown in Table 13.18.

TABLE 13.18: **Systemwise Allocation of Unavailability**

RAPS-1 (TOTAL 48955 HRS) 1973 TO MAY 1985			RAPS-2 (TOTAL 9550 HRS) 1980 TO MAY 1985		
Nuclear Systems:		74%	Nuclear Systems:		47%
End shield:	52%		Heat Transport	14%	
Heat Transport	18%		Moderator:	13%	
			Boilers:	8%	
			Instrumentation	7%	
Conventional Systems:		18%	Conventional Systems:		46%
Turbine:	15%		Turbine:	34%	
Grid Fault:		4%	Grid Fault:		5%
Miscellaneous:		4%	Miscellaneous:		2%

Major Equipment Problems

The following were the major equipment problems faced at RAPS:

Nuclear Power Systems

(a) *South End Shield RAPS-1:* (See Figs. 13.13, 13.14 and 13.15). (1981-1984) South end shield started leaking light water on September 20, 1981, from F-10 lattice position. As the leak was in a highly radioactive location (20,000 R/hr) with no access, first attempt was made to close the leak by chemical plugging (isopropyl alcohol and demin. Water was followed by zinc silicate solution circulation; fixing by ammonium hydroxide solution circulation and again isopropyl wash and DM wash). This helped to run the unit for 36 days but the leak reappeared on March 4, 1982. Next the fuel channels were removed from F-10, F-9 and E-10 and closure plugs with indium metal gaskets were installed in CSTS (Calandria Side Tube Sheet) and silver gasketed plugs in FSTS (Fuelling Machine Side Tube Sheet) bores in December 1984 successfully to stop the leak. This problem has cost the station nearly a 3-year shutdown. Repair was done using indigenous materials, men and methods.

(b) *Heat Exchanger Failures:* Heavy water/light water heat exchangers have shown up some leaks due to tube failures both in RAPS-1 and RAPS-2 as follows:

(i) Mod. Heat Exchanger No.1 - RAPS-2 (1976).
(ii) Mod. Heat Exchanger No.2 - RAPS-1 (1977) (1978).
(iii) Heat Transport Heat Exchanger No.1 - RAPS-2 (1979).
(iv) Heat Transport System Heat Exchanger No.1 - RAPS 1 (1982).

The leaky tubes were identified and plugged. Loss of heavy water and escape of radioactivity are the two dangers involved in such leaks though adequate instrumentation exists to warn the operator about the occurrence of the leak. Causes of the leak were established as:

(i) Flow induced vibrations.
(ii) Corrosion pitting due to stagnant water.
(iii) Erosion.
(iv) Fretting of tubes near baffies.
(v) Design/manufacturing defect.

All heavy water heat exchangers were subsequently eddy current tested as Inservice Inspection for predictive maintenance and tubes with more than certain percentage thinning were plugged to prevent leak. Operating practices have been improved and design reviews made.

Conventional Systems

(a) *Turbine Blade Failures*: In all there have been five incidents of blade failure, three in RAPS-1 and two in RAPS-2. Following are brief details:

(i) *First Failure RAPS-1, 4th Stage HP-July 1974:* One blade failed shearing from the root. 10 more blades of the 134 blades in this stage showed cracks. RAPS-2 rotor was temporarily used in RAPS-1 to enable start-up.
Cause: Excessive side grip clearance which resulted in resonance of blade vibrations in tangential mode very close to diaphragm impulse frequency.
Repair: Replaced with blades of new design. Original inverted 'T' root was changed to tapering root.

(ii) *Second failure RAPS-1, 3rd Stage HP-July 1975*: One blade had failed at the 'T' root as it happened in other rotor in 1974. 33 more blades showed cracks.

Cause: Same as in 1974.

Repair: As a temporary measure rotor was used without the third stage blades till safe design was received from suppliers. This third stage was subsequently rebladed in 1977 with blades of same design but natural frequency being kept clear of diaphragm Impulse Frequency.

(iii) *Third Failure RAPS-1, 5th Stage LP-Feburary 1976:* One blade failed. Random failure. 3 blades replaced.

(iv) *Fourth Failure, RAPS-2, 2nd Stage HP Rotor-January 1982:* Two blades sheared from the root.

Repair: As a temporary measure RAPS-1 rotor which had new modified blades in the 4th stage was installed with some reinforcement modifications. (Tack-welding of adjacent blades on both sides and understrapping of Stage-2 shrouds.) Damaged rotor sent to BHEL, Bhopal, for complete replacement of Stage-2 blades with new blades with lacing for rigidity. After repairs it waited to be installed in RAPS-1.

(v) *Fifth Failure, RAPS-2, 2nd Stage HP Rotor and January 1983*: Two blades had broken. Original Unit-2 rotor which was waiting in RAPS-1 was installed in Unit-2 and the unit commissioned within 25 days on February 21, 1983. The damaged rotor was sent to Bhopal for repairs and after repairs was used in RAPS-1.

Cause of Turbine Blade Failure: Abnormal frequency in the grid has been quoted by the manufacturers for the stresses leading to failure of the blades. The manufacturer has been requested to review the design in depth and supply spare rotor with blades so as to withstand grid frequency variations seen in the northern region.

(b) *Deaerator Problem (1973)-RAPS-1*: Water hammer and damage to trays were seen during large load throw-offs like turbine trips. This was caused by storage tank water flashing into steam, rising through equalizing area and tray stack and in a reverse direction the condensate flow coming from the top modifications were done by:

(i) Strengthening tray structure.

(ii) Change in control logic to reduce water flow on load throw off and deaerator level programmes to reduce set point on highei pressure.

(iii) Increasing equalizing area between deaerator and storage tank.

(iv) Spray nozzles in condensate distribution to prevent steam entry when flow is less.

(c) *Condenser-RAPS-2*: Tubes are of admirally brass. RAPS-1 performance is good whereas RAPS-2 condenser has given several leaky tubes – which were identified on-power and plugged.

(d) *Reheaters (1980, 1983, 1984) RAPS-2:* Reheater tube leakage has caused unit unavailability. One bank has been replaced while, in others, leaky tubes have been plugged. There is no spare capacity in the reheaters and a failure of any one of six in each unit affects the unit output.

(e) *Air Compressors (1973) RAPS-1*: Oil from compressed air used to find its way into valve positioners and actuators of safety equipment. Now oil free compressors have been installed.

Refuelling

Refuelling at RAPS is done using remotely operated fuelling Machines – two such required for each unit to work simultaneously one from each side of the reactor (south and north vaults). Refuelling is one of the biggest challenges at PHWRs as each refuelling involves opening up the high pressure, high temperature coolant channels by removing the seal plugs for inserting the fresh bundles and pushing out used bundles. The seal plugs have to be reinstalled in the coolant channels before the fuelling machines are retracted. Chances of heavy water spills always exist and call for quick alert operator by an action.

Fuel performance has been excellent. Of the 14,000 bundles loaded in the reactors, only 38 have failed (rupturing of cladding) so far.

Refuelling and fuel performance can be seen from Table 13.19.

On a few occasions when fuelling machines had problems while refuelling, it was necessary to shut down the unit and cool the system to retract the machine. But due to persistent request of grid authorities, unit shutdown was not taken and unit continued to operate for several days on the available reactivity only to find that a huge backlog developed in refuelling. Incidents were as follows:

TABLE 13.19: **RAPS Refuelling and Fuel Performance (1975 to May 31 1985)**

	RAPS-1	*RAPS-2*
1. Maximum Power level	212	212
2. Full Power Days Cumulative	1393	880
3. Total No. of channels refuelled	1219	749
4. Total No. of bundles discharged	8981	5492
5. Failed fuel bundles	23	15
6. Average fuel burn-up of discharged bundles (MWD/TeU)	6700	7200

RAPS-1:

(1) 23.9.80 to 13.10.80 on E-18 –20 days.
(2) 13.2.81 to 22.03.81 on P-4 –40 days.

Refuelling has never been able to keep up with of the unit operation. Refuelling in RAPS-1 started on October 20, 1974, at 174 FPD, 21 FPD later than required. In RAPS-2 refuelling started on November 26, 1981, at 174 FPD, 50 FPD later than required. Constant backlog in refuelling has at times resulted in derating of the units.

Heavy Water Management

Heavy water is used as coolant and moderator in PHWR: Each unit needs about 250 tonnes as initial requirement and about 15 to 20 tonnes per year as make-up for losses. In the heat transport system about 60 tonnes are required. Majority of this is at 87 kg/cm^2 and 260-290°C when the unit is operating. Because of the number of valves, sealing plugs, pumps in the system, some heavy water always escapes into the reactor building. Reactor grade heavy water has to have purity of 99.8 per cent and above. What comes out of the process systems into the building is called "Escapes".

Recoveries

What escapes into the reactor building is remotely recovered using vapour recovery driers. These are called Recoveries. Because there also exist light water circuits in the reactor building, these recoveries get downgraded in concentrations. Because of D20 scarcity, even 0.3 per cent isotopic degraded collections are preserved for upgrading.

Losses

What is not recovered from the escaped heavy water from reactor building finds its way to the stack and atmosphere as loss.

Upgrading Plants

There are distillation columns. The first one at RAPS was imported. Now there are three additional indigenous upgraders. Two of them have already been commissioned and are operating satisfactorily.

Strategy for Heavy Water Management

This comprises keeping the losses low and recoveries high and prompt upgrading of the recovered water so that it is available for recycling without depending on fresh supplies which are *SCARCE*.

Heavy water Performance

Typical figures from heavy water accounting at RAPS are as under:

Escape Rates	:	150-200 kg/day/reactor.
Recovery Rates	:	135-170 kg/day/reactor.
Loss Rates	:	15-30 kg/day/reactor.

Heavy water performance in RAPS-2 has been better than in RAPS-1 because of better leak tightness of boiler room atmosphere from accessible areas in RAPS-2.

Demand on the Grid

A nuclear power station needs the following from the grid:

(i) Uninterrupted power supply to over-ride poison and avoid poison outs.
(ii) Independent power supply sources for redundancy.
(iii) Voltage and frequency within specified limits, say, 49.5-50.5 cps and 205-240 KV.

High voltage affects transformers and control power supplies. Low voltage trips motors on overload. Turbine manufacturers specify turbine operation above 49.5 cps otherwise turbine bearing vibration problems appear. Over-frequency and under-frequency affect reactor coolant system flows and indirectly affect reactor regulatory systems.

Grid should be at least 10 times as large as the highest rated generating unit connected to it. Then only it can sustain a unit outage. When RAPS-1 was first started up, it was a major portion of Chambal grid (500 MWe), almost 30 per cent; every time RAPS-1 tripped, the grid collapsed. This was remedied in January 1975 by paralleling Chambal grid with BBMB grid/Northern grid (5000 MWe) and grid stability improved.

RAPS was at the tail-end of the transmission system with all other stations located away on one side. High voltages under low load conditions and low voltages under high load conditions were very common at RAPS. A ring main would perhaps solve this problem. (in Fig. 13.16).

Droop Settings

All hydro and thermal stations were keeping their droop settings high. RAPS droop is 4 per cent and was getting all the fluctuations. Hydro and thermal settings were brought down to 3.5 per cent and 4 per cent which has helped RAPS.

Unit Islanding

On low frequency RAPS separates from Kota and continues on Udaipur load on islanded mode till Kota (Chambal-BBMB) picks up load. Because of a possibility of poisoning out with half hour, Udaipur load should be at least 70 per cent of the maximum operating power of RAPS.

Feed-back from RAPS Operating Experience

Being the first PHWR type reactor in India, RAPS operating experience since 1973 has been very valuable for DAE Design Engineers to make improvements in subsequent projects at MAPP and NAPP. Some of the salient points are:

(a) *Containment Building:* Double walled containment at MAPP and NAPP in view of higher leak rate at RAPS which had single wall containment,

(b) *Dousing Eliminated:* RAPS dousing system has been eliminated in MAPP and NAPP and other projects by replacing with vapour suppression pools,

(c) *Closed Loop for Reactor D20 System:* Consequent on D20 heat exchanger leak incidents, all Reactor Building D20 heat exchangers are cooled by closed-loop cooling circuits not directly connected to lake or river streams,

(d) *Refuelling Machine Logic Circuits and Equipment*: Improved for better maintainability,

(e) *Moderator Dumping Eliminated:* Moderator dumping has been eliminated from NAPP onwards. This will help to decrease poison shutdowns,

(f) *End Shield Material Improved:* Stainless steel is now used for end shields from MAPP-2 onwards,

(g) *Dump Condenser on Boiler Steam:* These are provided from MAPS-1 onwards. This will help dimineralized water conservation and poison override,

(h) *On-load Tap Changers for Start-up and Unit Transformers:* These are provided from MAPS-1 onwards for control of voltages while on-power, and

(i) *Heavy Water Management*

(i) Commissioning of D20 system using light water to reduce waiting time and initial D20 losses was started with RAPS-1,
(ii) Reduction of D20 valves,
(iii) Better sealing of atmospheres,
(iv) Use of below seal valves,
(v) Regular inservice inspection of D20 Heat Exchangers by Eddy Current problem of tubes and remedial repair, and
(vi) *Separation of H20 and D20 Systems*: NAPP boilers have been located separate from D20 circuit equipment so that downgrading of heavy water is minimum.

(j) *Grid Interface*

(i) Paralleling of sub-grid with main grid.
(ii) Islanding.

MAPS EXPERIENCE

General Description

Located at Kalpakkam, 80 kms from Madras, MAPP is the second PHWR but the first PHWR to be designed, manufactured, built, commissioned and operated with 100 per cent Indian contribution. It has 2 × 750 MW(TH) (235 MWe) units similar to RAPS.

In a way, MAPS-1 marks the coming of age of the Indian atomic energy programme as it gives a proof of complete self-reliance in nuclear power plant capability in all its inputs. An indoor switchyard to overcome the saline effects of the sea winds and a submarine tunnel to overcome the silt problem are two items done for the first time in the country.

Operating Factors, Capacity Factors and Outages

The performance of MAPS-1 has been better than the imported unit RAPS-1 in its comparative period of operation. The first year of commercial operation January 27, 1984, to end December 1984 has shown very encouraging results for the first Indian PHWR unit in Table 13.20.

TABLE 13.20: **MAPS-1 Operating and Capacity Factors and Outages (1983-1984)**

Period	Operating Factor (%)	Capacity Factor (%)	Outages
23.7.1983 to 31.12.1983	43.2	23.5	22
27.1.1984 to 31.12.1984	75.3	64.3	24

Outages

The number of outages in 1983 were 22 while in 1984 they were 24 in number.

Major Causes of Outages and Unavailability

(a) In 1983 major causes were:

(i) Reactor side 10 out of 22.
(ii) Conventional 12 out of 22.

(b) *In 1984:*

(i) Reactor side 9 out of 24.
(ii) Conventional side 15 out of 24

(c) Systemwise significant causes are given in Table 13.21.

TABLE 13.21: **Major Causes of Unavailability MAPS-1**

	Reactor Side		Conventional Side
(i)	Steam leak from boiler manhole cover	(i)	BFP problem
(ii)	Boiler level control	(ii)	Reheater drain tank steam leak
(iii)	Adjuster rod low flow	(iii)	Blockage of sea water screen by jelly fish
(iv)	PHT pump bearing oil draining	(iv)	Condenser low vacuum
		(v)	Turbine high vibrations

Major Equipment Problems

Reactor Systems

(a) Thermal shield fan high vibrations,
(b) Blockage in T-8, T-10 and S-7 feeder pipes were subsequently cut and blockages were removed and feeder pipes were normalized,
(c) Biological shield cooling pipe leak. Remedied by chemically plugging,
(d) Channel J-20 had to be replaced, and
(e) Boiler manhole cover steam leaks.

Conventional Systems

(a) Turbine bearing No.5 had high vibrations remedied partly by stiffening the plate of bearing casing and by trim balancing. Problem still exists,
(b) Boiler feed pump high vibrations. Supports strengthened but

still need improvement, and

(c) Expansion bellows of HP exhaust leaked so also the inlet to PHS reheater. Subsequently expansion bellow was replaced.

Refuelling

(a) Because of use of 656 depleted uranium bundles, while initial fuel loading, sufficient flux flattening could be attained and reactor power could be raised to 95 - 100 per cent Full Power by May 1984

(b) On-power refuelling was started at 125 EFPD (18.5.84). Fuel management was done using TRIVENI computer programme. Totally 242 channels were refuelled by December 1984 and 1,222 fuel bundles were discharged to storage bay. 722 bundles were recycled. Maximum burn-up achieved is 8,400 MWD/TeU and average about 4,345 MWD/TeU.

Heavy Water Management

Upgrading effort has been excellent at MAPS. The indigenous electrolysers operated 88 per cent of the time in 1983 and processed 480 tonnes of 6 per cent to 90 per cent isotopic to produce 52 tonnes of reactor grade heavy water. In 1984 similar performance was repeated.

Heavy water escape rates in 1983 and 1984 were brought down from 125 kg/day to 100 kg/day from PHT system. Stackloss figures are less than 20 kg/day. These have resulted from the improved sealing of boiler room atmosphere.

Feed-back from MAPS-1 Experience

The following are the significant feed-backs:

(a) Commissioning of turbine and generator using auxiliary steam boilers well ahead of nuclear steam availability to check out turbine and generator was tried the first time at MAPS-1 very successfully. MAPS-2 turbine and generator have already been tested at speed similarly,

(b) Reactor building exhaust fans have been provided with two-step speed control: Low speed during reactor operation to reduce atmospheric moisture ingress in Boiler Room and high speed during maintenance work to reduce air activity levels.

Radiation Safety and Environmental Pollution

Atomic Energy Act 1962 provides for control over radiation generating units to:

(i) prevent radiation hazards,
(ii) secure public safety and safety of radiation workers,
(iii) ensure safe disposal of wastes.

Radiation Protection Rules have been promulgated (1971). Health Physics Division of BARC, Division of Radiological Protection, BARC provide the surveillance for monitoring the radiation fields, radiation dose. These bodies are independent of the power stations management administratively.

Environmental Survey Labs have been set out, one at each power station, to monitor the benchmark data before start of reactor operations and then on regularly to monitor the subsequent data to check the effect of nuclear power operations on the environment.

Radiation Protection Training is imparted to each worker and he is checked out to test his understanding. Use of protective clothing, prior planning of work and initial practice at "mock-up" of the equipment and methods in advance help to reduce Manrem Exposure.

The record of radioactive effluents into the environment as also the radiation exposure of personnel at each power station is kept by the Health Physics Division posted at the station. Table 13.22 shows what the records so far indicate.

TABLE 13.22: **Manrem Exposure and Gaseous Effluents**

Stations	Average Annual Exposure	Gaseous Effluents
TAPS	1.2 Rem (1969 - 1983)	10% of Permissible
RAPS	Less than 1 Rem (1973 - 1983)	10% of Permissible (Argon)
MAPS	0.3 Rem (1984)	3% of Permissible

At RAPS, the Solar Evaporation Facility installed and commissioned since 1976 has totally eliminated any discharges of active waste into the lake. Even when they were earlier discharged into the lake it was with sufficient dilution as per International Standards so as not to effect aquatic life and flora and fauna in the region. TLD monitors placed in the different locations and samples analysed at ESL prove that the effect on the public and environment in the vicinity of the station and up to 60 kilometres is within permissible levels or almost insignificant considering natural background values.

Concluding Remarks and Prospects

The operational experience gained at TAPS, RAPS and MAPS indicates that these nuclear power stations have proved their technical and economic viability in the Indian grid. With each new nuclear power plant commissioned, there is clear evidence of overcoming the problems faced in the earlier plants.

The long three-year shutdown of RAPS-1 (1982-84) to repair the south end shield leak might raise some doubts among the operating engineers regarding the long-term operability of such units. RAPS-1 (1964 design) is essentially a proto-type in PHWR and many later versions to overcome such problems have since been built and commissioned. The spectacular progress of nuclear power plants, however, has not always been smooth throughout. Far from dampening the spirit, such problems have in fact spurred the nuclear engineers and scientists to face them with increasing challenge.

In the Indian situation, there are clear signs of the confidence among the engineers and scientists of the DAE in delivering the targeted nuclear generating capacity in the decade and half ahead.

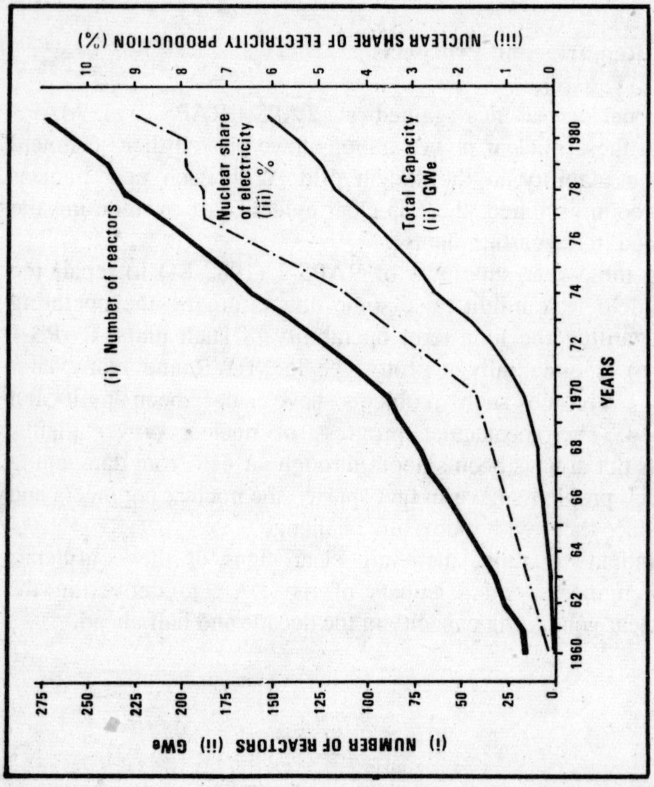

Fig. 13.1: The Development of Nuclear Power in IAEA Member States, 1960-81.

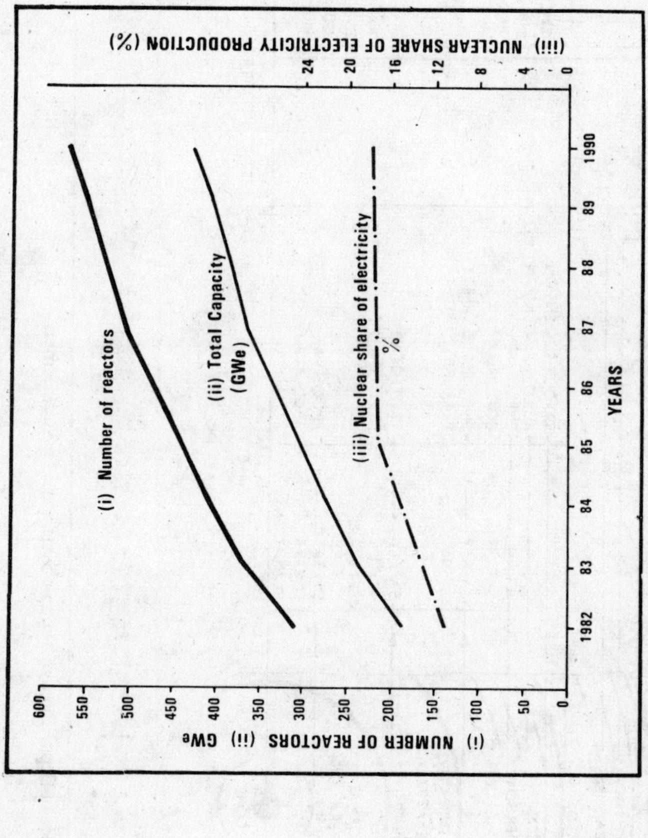

Fig. 13.2: Projected Growth of Nuclear Power in IAEA Member States 1982-1990.

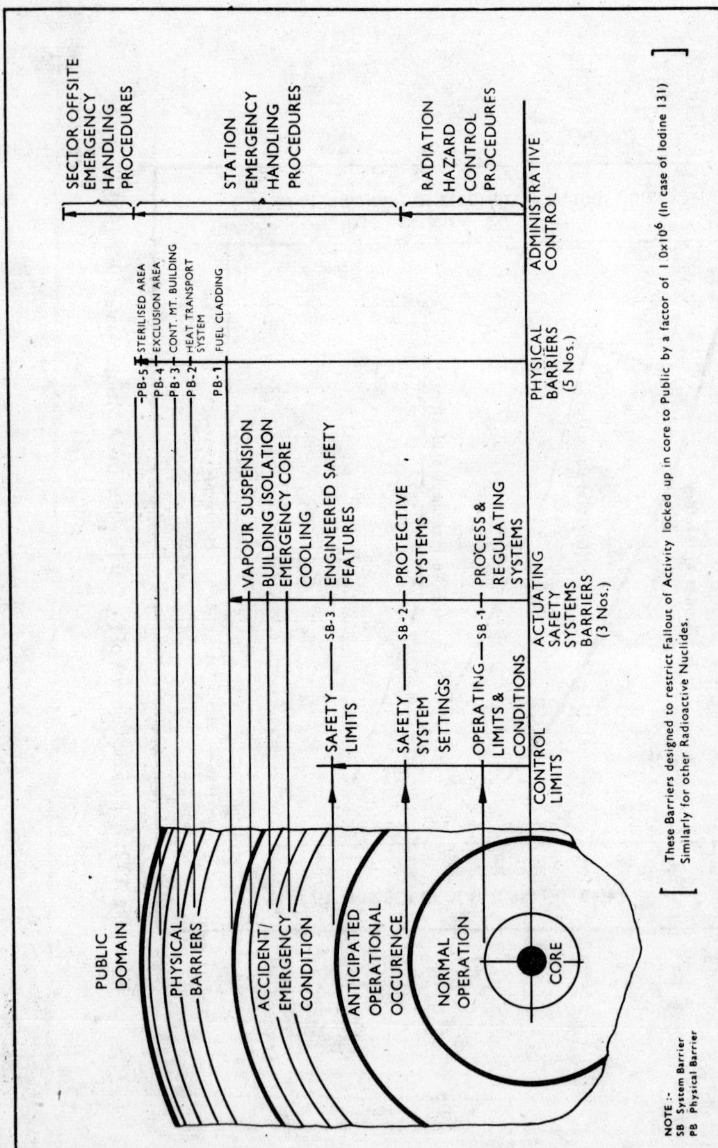

Fig. 13.3: Defence in Depth Through Multiple System and Physical Barriers.

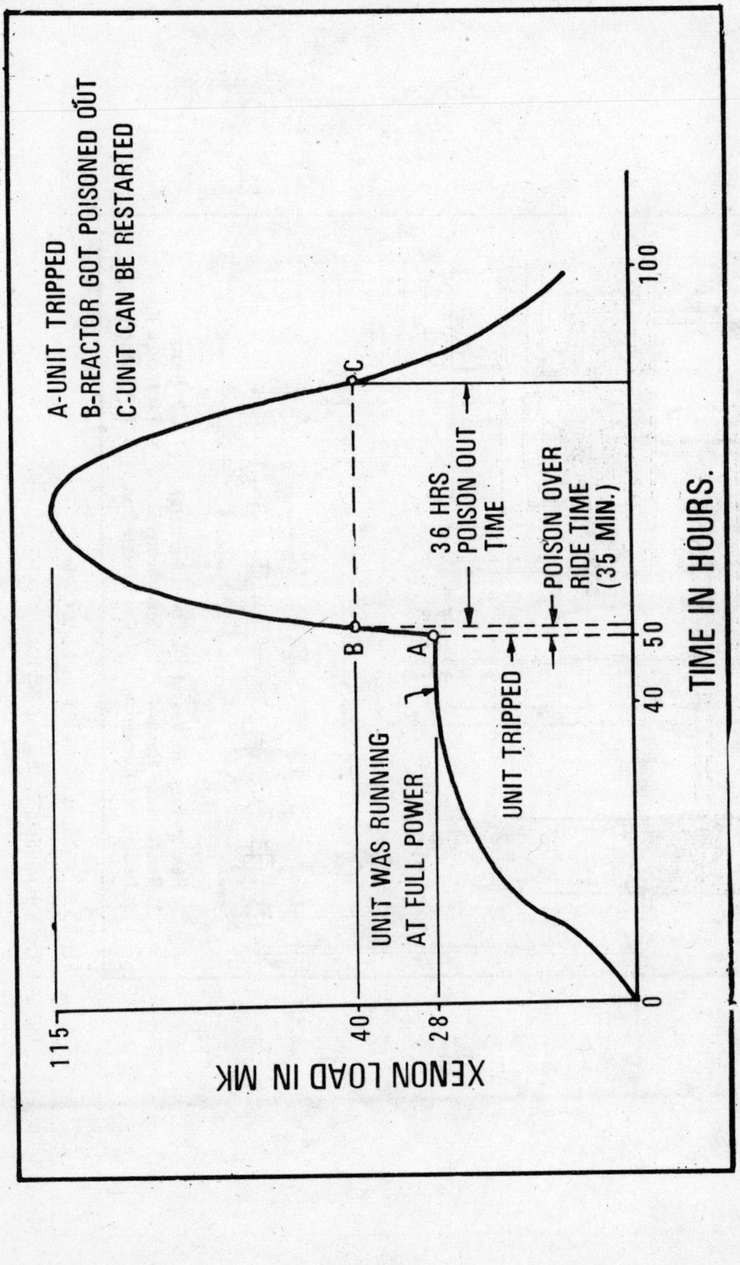

Fig. 13.4: RAPS Xenon Load Curve During Operation and after Tripping the Reactor.

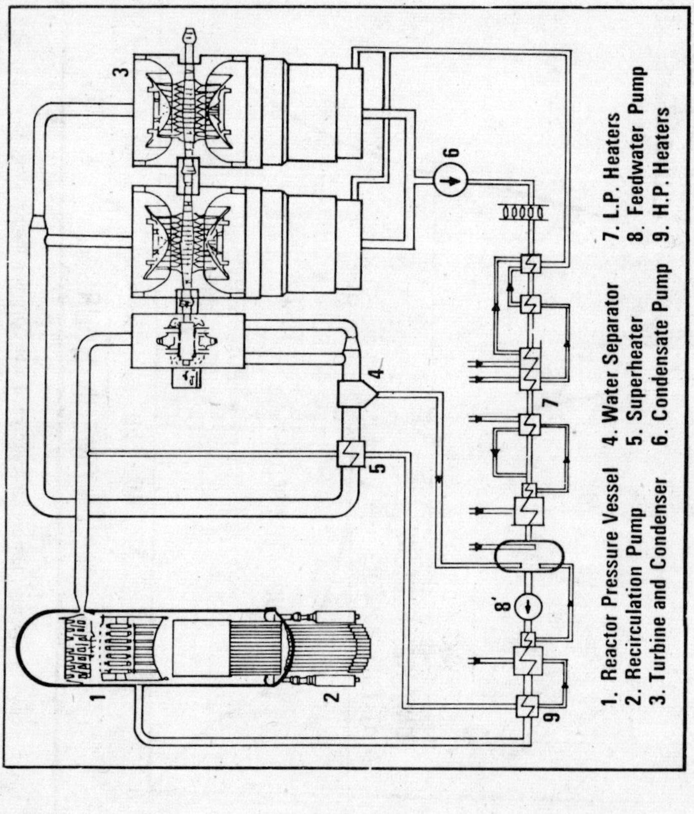

Fig. 13.5: Boiling Water Reactor Main Cycle.

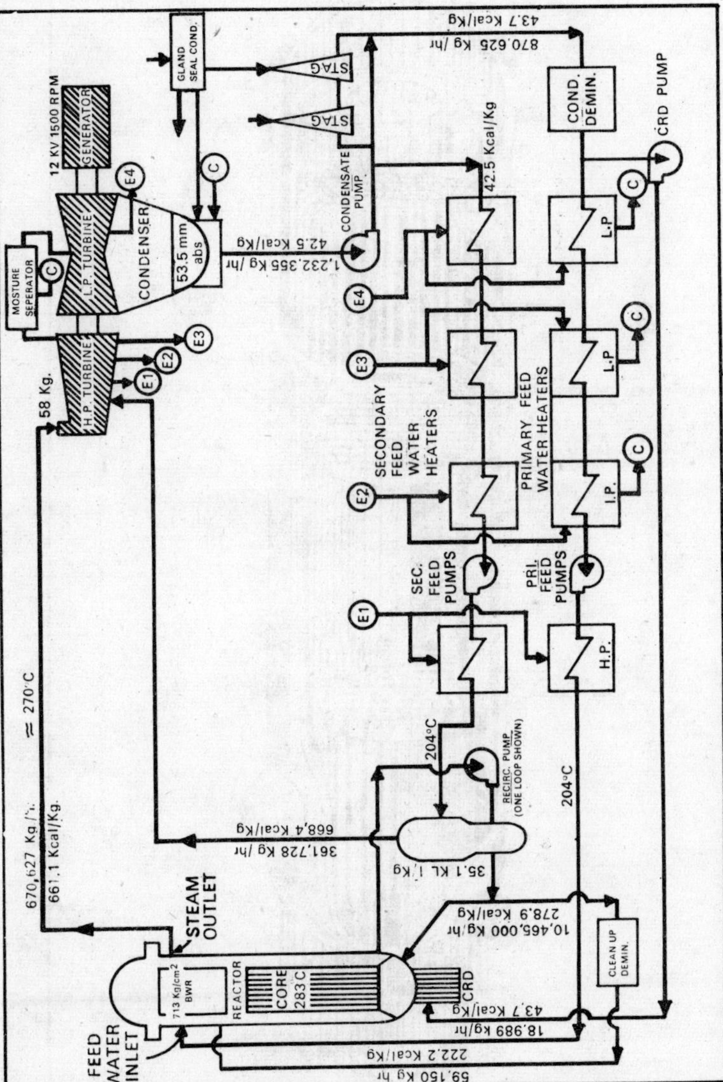

Fig. 13.6: TAPS Diagram of System and Feed Cycle.

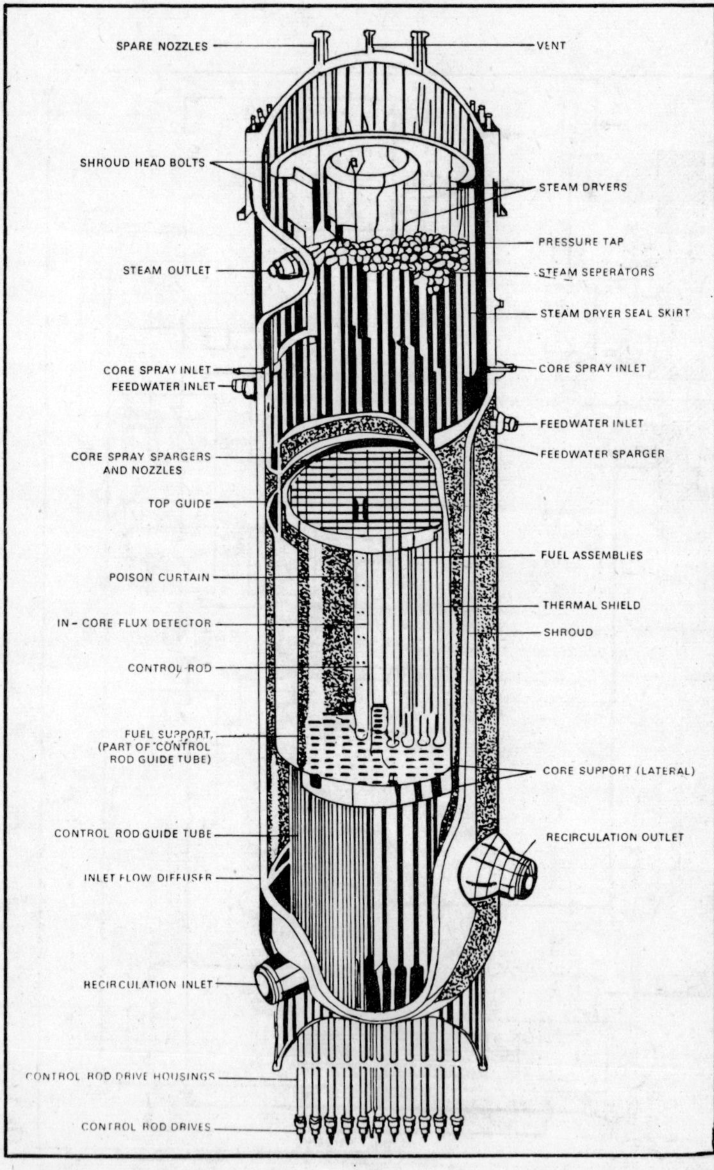

Fig. 13.7: TAPS Reactor Vessel

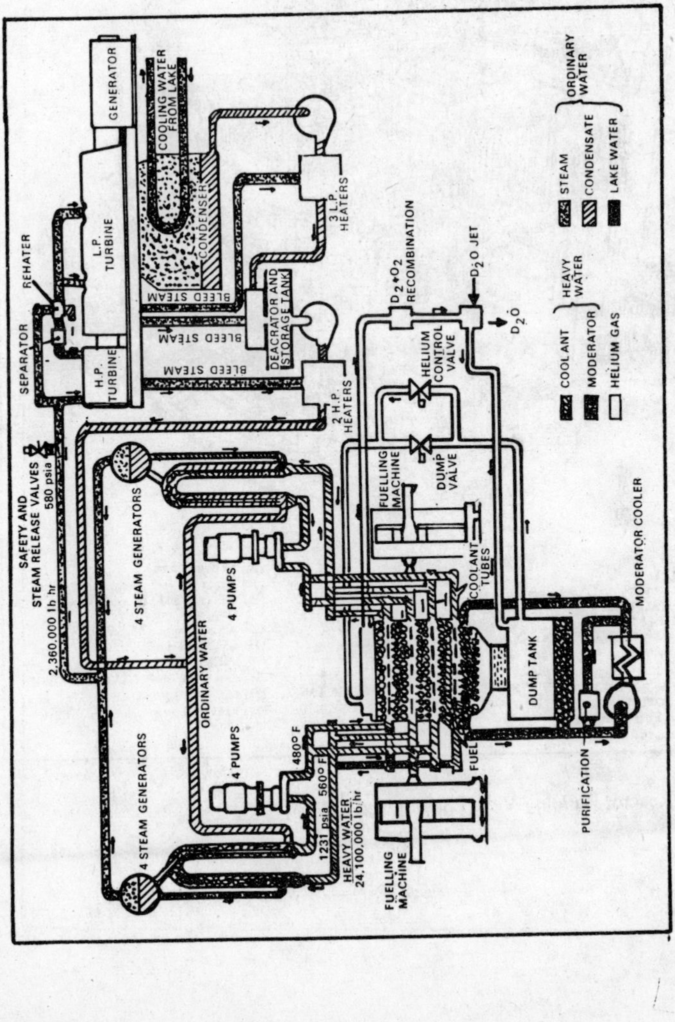

Fig. 13.8: Simplified Flow Diagram of Rajasthan Atomic Power Station.

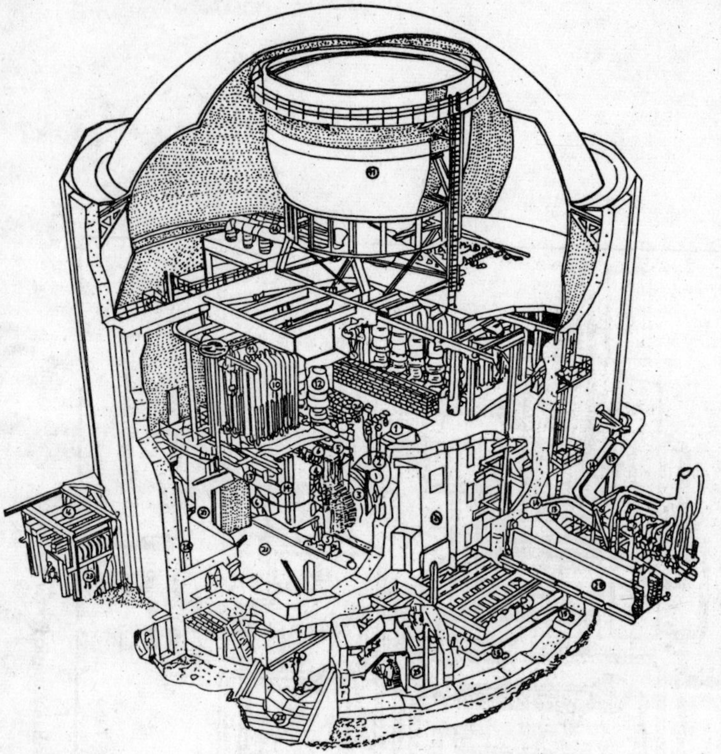

(1) CALANDRIA
(2) HELIUM LINE
(3) REACTOR SHIELDS
(4) PRIMARY HEAT-TRANSPORT SYSTEM FEEDERS
(5) FUELLING MACHINE
(6) CABLE CART
(7) BOOSTER RODS
(8) PRIMARY HEAT TRANSPORT REACTOR OUTLET HEADER
(9) PRIMARY HEAT TRANSPORT REACTOR INLET HEADER
(10) STEAM GENERATORS
(11) STEAM DRUM
(12) PRIMARY HEAT TRANSPORT SYSTEM MAIN PUMPS
(13) STEAM HEADERS
(14) BOILER FEEDWATER HEADERS
(15) PROCESS WATER INLETS
(16) VENTILATION DUCTS.
(17) EMERGENCY WATER STORAGE TANK
(18) SPRAY TANK PIPES
(20) CABLE TRAYS
(21) FUELLING MACHINE VAULT
(22) SPENT RESIN STORAGE TANK CHAMBER
(23) AIR LOCK
(24) LABYRINTH
(25) ELEVATOR
(26) THERMAL SHIELD COOLING DUCTS
(27) BLOWOUT PANEL
(28) PRESSURE WALL

Fig. 13.9: Reactor Building View Looking South West.

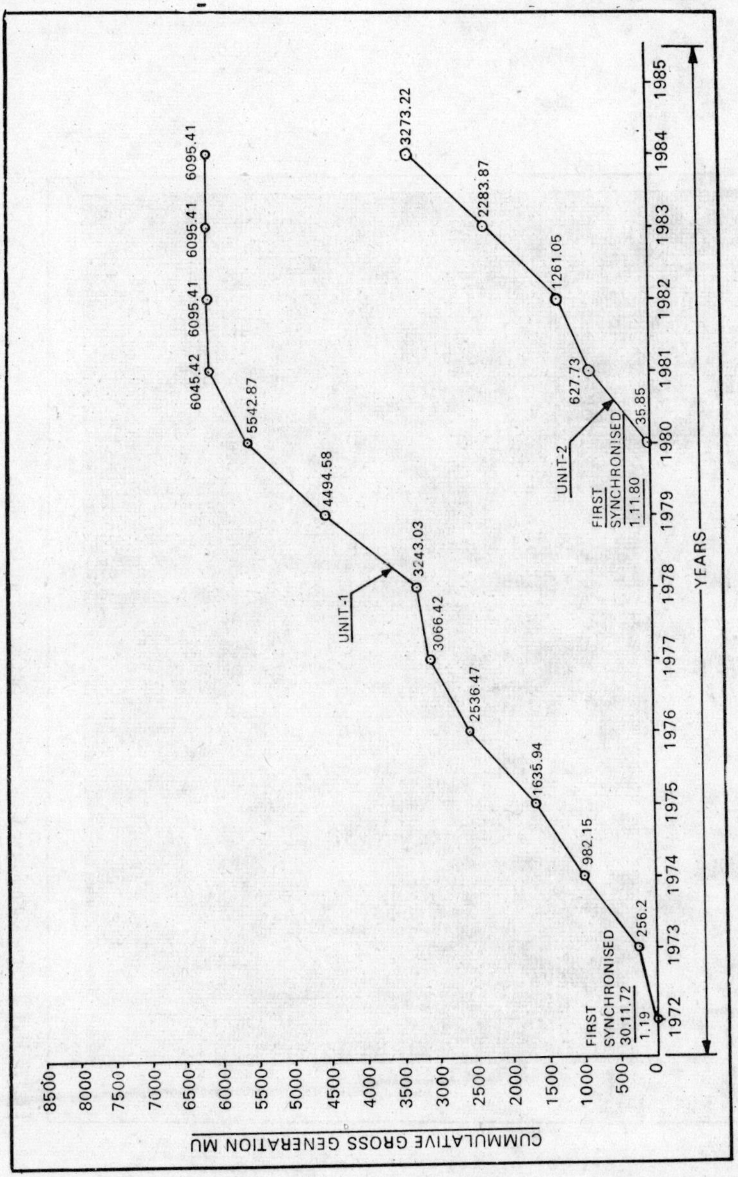

Fig. 13.10: Cumulative Gross Generation of RAPS

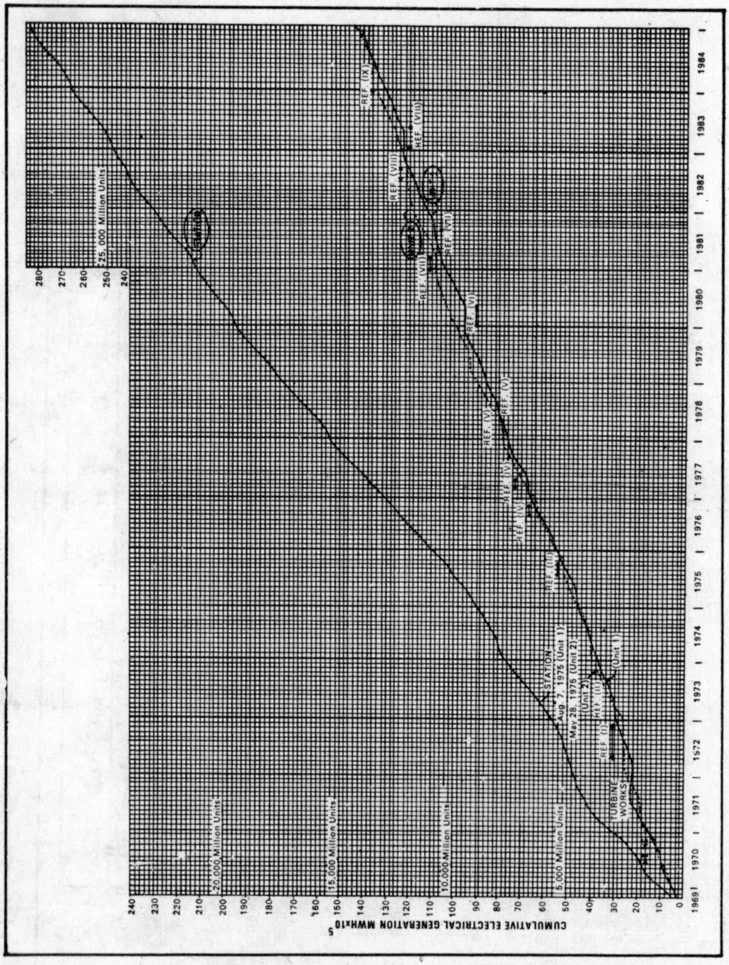

Fig. 13.11: Tarapur Cumulative Generation Since Commercial Operation.

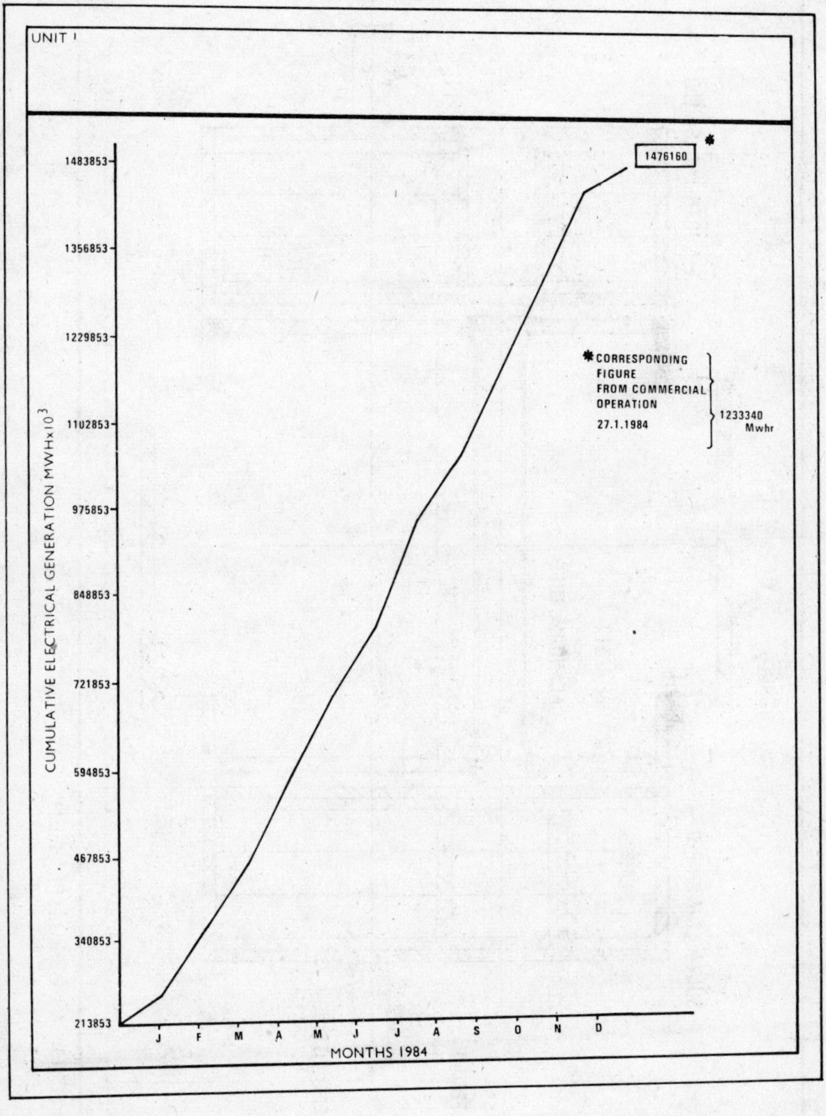

Fig. 13.12: Madras Atomic Power Station — Cumulative Generation in MWhs.

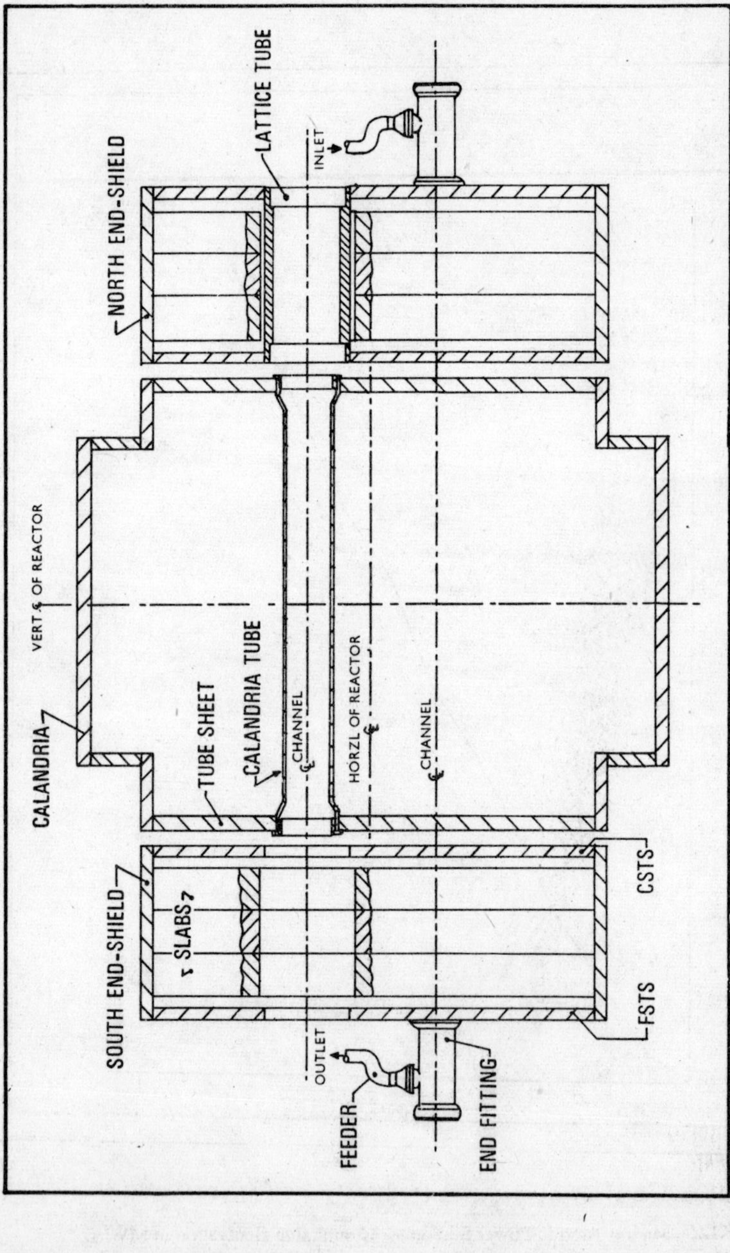

Fig. 13.13: Calandria and End Shield.

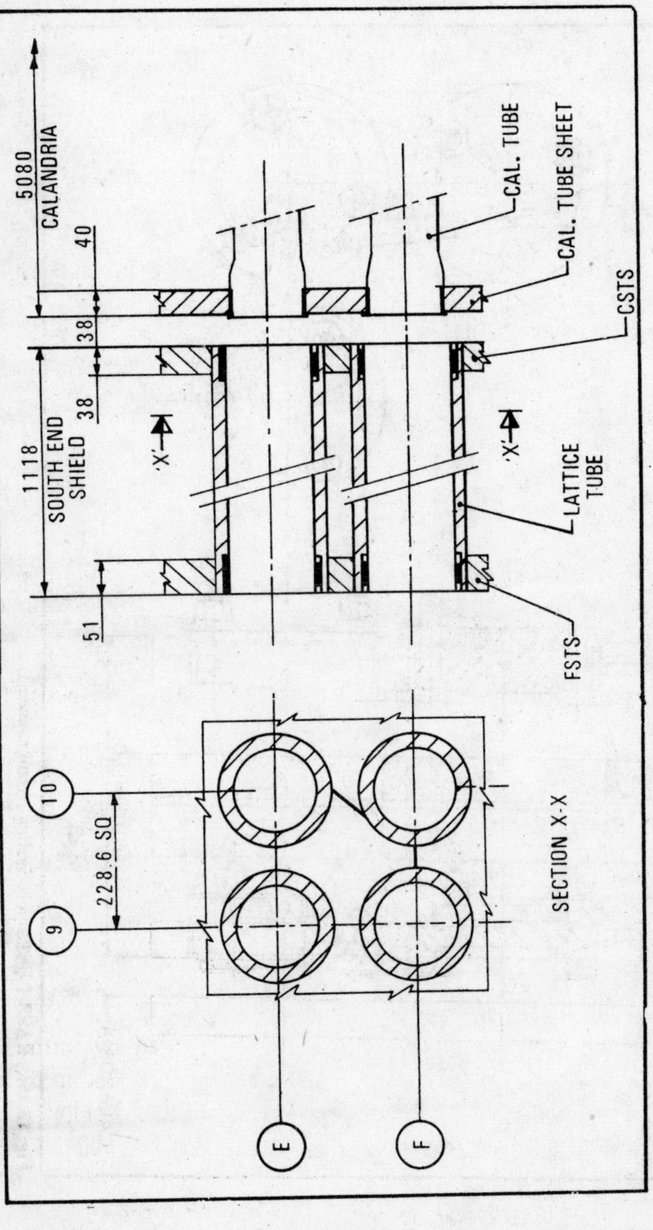

Fig. 13.14: Location of Cracks in Calandria Side Tube Sheet.

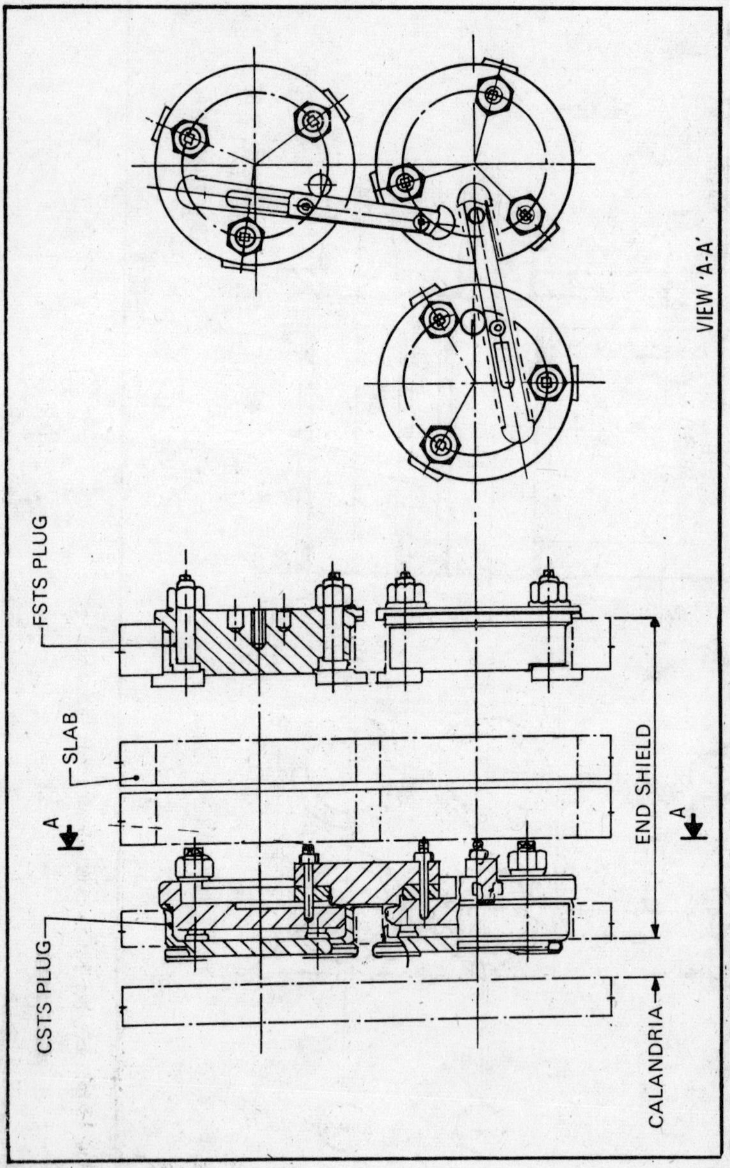

Fig. 13.15: RAPS-1 End Shield Sealing Arrangement.

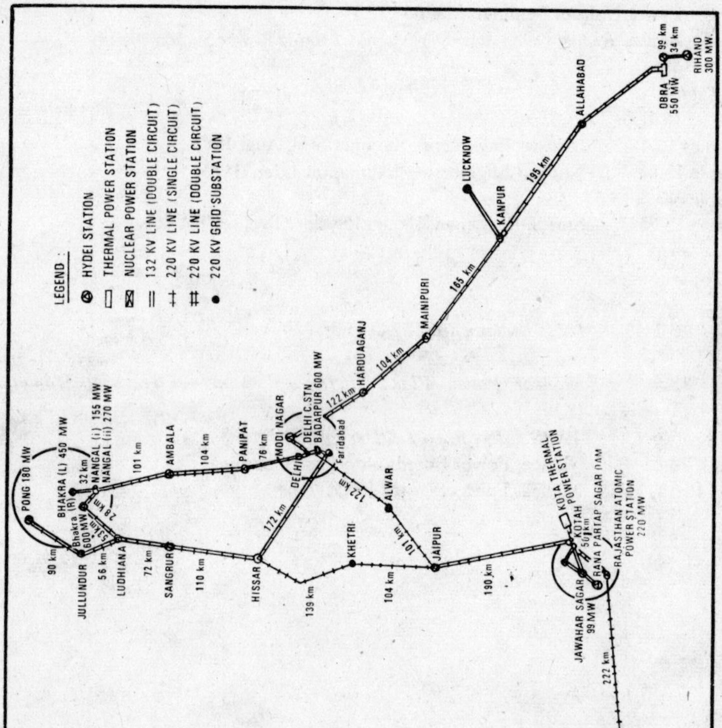

Fig. 13.16: Part of Northern Grid.

REFERENCES AND ACKNOWLEDGEMENTS

1. *General*

1. *Performance Report* (1984), Tarapur Atomic Power Station.
2. *Nuclear Power Stations O&M Experience* by Shri K.P. Rao and Shri M.S.S. Raghavan, National Seminar on Power Stations, 6.1.1984, Madras.
3. *Annual Report (1984),* Rajasthan Atomic Power Station.
4. *O&M Experience of RAPS* by Shri M. S. R. Sarma, National Council of Power Utilities Seminar, Madras.
5. *Annual Report* (1983 & 1984), Madras Atomic Power Station.

2. *Tables*

Tables 13.1-13.2	Nuclear Engineering International, Aug. 1983.
Table 13.4	Nuclear Engineering International, Jan. 1985
Tables 13.5-13.6	" " " " " "
Table 13.21	Shri S. D. Soman, Health Physics Division, BARC.

3. *Figures*

Figs. 13.1-13.2	*IAEA Bulletin* 1982 H. J. Laue.
	"
Figs. 13.5	*Nuclear Power Plant Operational Safety — Training Manual* (KWU) 1981.
Figs. 13.6-13.7	TAPS *Safety Report & Design.*
Figs. 13.13, 13.14, 13.15	Nuclear Power Board (Indo-French Seminar on Nuclear Energy) (April 1985) *Papers.*

FOURTEEN

FABRICATION OF NUCLEAR COMPONENTS

A. KRISHNAN

INTRODUCTION

During recent years India has made a rapid progress in implementation of the Nuclear Power Programme. In this one of the main hurdles experienced had been the long gestation period required for building nuclear reactors. This was partly because of lengthy manufacturing schedules of critical core components. Difficulties were faced in early completion of fabrication of these components because of lack of experience on the part of Indian manufacturers in meeting the stringent nuclear quality standards. The reactor components like Calandria and End Shields are fabricated from austenetic stainless steel (Grade 304 L) and the Indian manufacturers did not possess enough experience in the fabrication of stainless steel components of this nature.

However, considerable progress has since been made, and now we can reduce the overall fabrication time while meeting stringent quality and safety requirements. This is expected to shorten the gestation period and make nuclear power cheaper (or more competitive, etc.)

This paper briefly describes the breakthrough achieved in the manufacture of the main reactor vessel called Calandria where the effective completion time of fabrication has been reduced from seven years to one year. Fig. 14.1 shows the Integral Calandria/End shield assembly.

Welding

Manufacturing ability of an industry primarily depends on its welding and machining potential and its capability of feeding the required

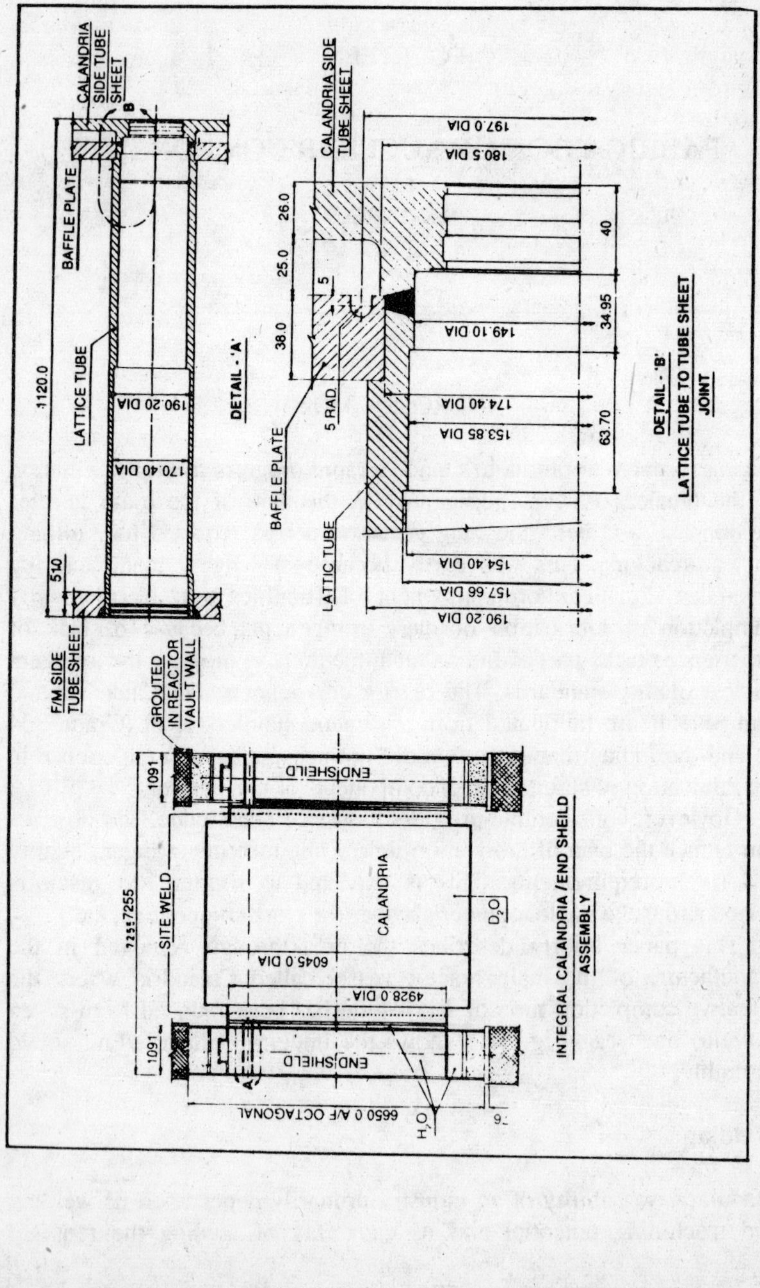

Fig. 14.1: Integral Calandria/End Shield assembly.

engineering know-how to its different shops. There has been continuous progress in all spheres during the course of indigenous manufacture of major stainless steel reactor vault components.

Stainless steel material, as we all know, possesses the requisite qualities and is an appropriate choice for fabrication of reactor components for our pressurized heavy water reactors. But at the same time, one has to equip oneself to tackle problems such as large welding distortions, machining problems, limitation of ultrasonic examination of materials and weldments and strict requirements of cleanliness rendering stainless steel to impart its full corrosion resistance potential.

To minimize distortions, the following guidelines have been followed during fabrication of components:

(i) The shrinkage forces to be kept as low as possible by using only that amount of weld as is required,
(ii) Fewer passes with large electrodes are preferable to a greater number of passes with small electrodes,
(iii) Placing the welds near the neutral axis would reduce distortion by providing a smaller leverage for the shrinkage forces to pull the plate out of alignment,
(iv) When the welds are placed around the centre of gravity, one shrinkage force would balance against another,
(v) Employment of back step welding or planned wandering or block welding technique,
(vi) By giving initial preset to the parts to be welded,
(vii) Adoption of a well-planned welding force against each other,
(viii) Reducing heat input by choosing lower amperage, higher welding speed and use of low heat input process,
(ix) Forcibly restraining the parts being joined from movement during welding by suitable jigs and fixtures.

In the field of weld technology, manual metal arc (MMA) and TIG techniques still remain the prime mode of welding in the fabrication of major nuclear components.

In order to further improve the fabrication quality and to save fabrication time, use of pulsed MIG, pulsed TIG, SAW, etc., are also being used.

Machining problems

As regards machining, both public and private sectors within the country are today well equipped to undertake even complex machining jobs. The major machining problems arise from the inherent properties of the material which render them more difficult to machine than, say, carbon steel. The marked spread between UTS and yield strength causes "gamminess" in machining which results in build-up on tool edges, poor finish, excessive heat and short tool life. Proper chip curler tools and machining parameters such as feed, speed and depth of cut were, therefore, evaluated by carrying out trials on mock-ups.

Quality Control of Austenetis Stainless Steel Materials and Welds

The quality control of raw materials and welds is very stringent for our nuclear components due to two reasons:

(a) Limited in-service inspection facilities due to inaccessibility during operations.
(b) To have a trouble-free equipment for the entire period of the reactor life without any accidents which may give rise to release of radioactivity.

The following NDT methods are being followed to achieve the above quality:

(i) *Dye-penetrant Inspection:* Dye check is done at different stages from the root onwards depending upon the criticality of the weld and non-accessibility for other NDT processes. In some cases, each pass is dye-checked (e.g. Tri-junction joint) and all weld edges are dye-checked. The limitations of dye-check are that it indicates only surface defects and not sub-surface defects; hence dye-check is always coupled with radiography or ultrasonic testing. The dyes used are checked for hallide content before use on stainless steel to prevent stress corrosion cracking in service,
(ii) *Radiography:* Radiography sensitivity normally specified is 2 per cent of the thickness and 100 per cent radiography for all nuclear components,
(iii) *Ultrasonic Testing:* All plates and forgings are ultrasonically tested both by normal and shear wave. Ultrasonics is also used for inspection of many joints in end shields and wherever radiography is not possible,

The limitation of UT in austenetic stainless steel is basically due to the grain size and structure. Large grain size produce an attenuation of ultrasonic by scattering of the directed sound wave and effectively reduces the strength of reflected signals which results in difficulty in detection by ultrasonics.

Cleanliness Requirement

Stainless steel exhibits excellent corrosion resistance by forming a thin, continuous adherent film of chromium oxide. The cleanliness of stainless steel is very important at all stages of fabrication in order to maintain this corrosion resistant film.

The problems which arise due to not maintaining standard of cleanliness are:

(a) *Chloride Contamination of Stainless Steel:* The main source of chloride contamination are masking tapes, grease, oil, etc. If stainless steel is contaminated with chloride during fabrication, it must be passivated to avoid pitting and stress corrosion cracking later in service. This contamination can be confirmed by chloride contamination test,

(b) *Iron Contamination of Stainless Steel:* This type of contamination is observed when stainless steel comes in contact with carbon steel and other allied products and materials and the result is pitting type of corrosion. Iron contamination of stainless steel can be established by "Copper Sulphate Test". This can be avoided by working in Dust Free Enclosure and using poison plates between stainless steel and carbon steel. The contamination, if loosely bound, can be removed by wire brushing and if these are embedded, then pickling is the only solution.

Steps Taken to cut down Fabrication Time of KAPP Calandria

NAPP/KAPP Calandria design is different from RAPS/MAPS. It is a cylindrical shell with both ends open having 77 nozzles as shown in Fig. 14.2. The concentricity requirement between free end and root of the nozzle is 0.1 mm. To maintain this, distortion of the nozzles during welding has to be controlled. The following sequence was adopted for KAPP calandria to reduce the fabrication time of calandria from seven years to one year:

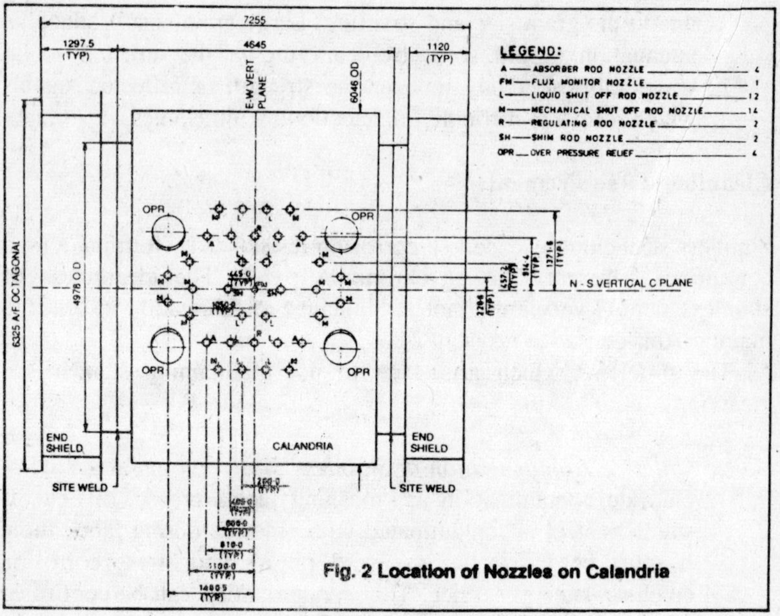

Fig. 2 Location of Nozzles on Calandria

Fig. 14.2: Locations of Nozzles on Calandria.

(i) Fabrication sequence changed, i.e., main shell and sub-assembly of small shell and annulus plate (HAT) were made in parallel. With this, flatness requirement of annulus plate was met (refer Fig. 14.3(a),

(ii) Nozzle openings were plasma cut (10 mm undersize) and then machining was done to make weld-edge preparation,

(iii) Final machining of all radial nozzles and rough machining of non-radial nozzles were done prior to welding with main shell (refer Fig. 14.3(b),

(iv) Nozzles to main shell welding was done in three stages, i.e.,

(a) welding of all radial nozzles,

(b) welding of 1st stage non-radial nozzles excluding threaded nozzles,

(c) welding of 2nd stage non-radial nozzles (threaded nozzles).

(v) Optimization of weld joint geometry to give adequate access to the welder as well as keeping in view of NDT requirements,

(vi) Simplest position, namely, flat (IG) was adopted for welding of girth joint and nozzle to shell joints,

(vii) Flattened GTAW inserts of size 1 mm x 2 mm were used to achieve 100 per cent penetration at the root in place of bigger insert,

(viii) Temporary stiffener rings were added to the main shell at four locations to maintain ovality within limits, during nozzle welding,

(ix) Proper distortion control fixtures and heat sinks were used during nozzle-to-shell welding.

By adopting the above, repairs in welds were negligible. These worked out to hardly 2 per cent of radiography spots numbering about 700.

Conclusion

By keeping the weld repairs minimum and by following the above guidelines, fabrication time of critical components like Calandria and End shields can be reduced considerably.

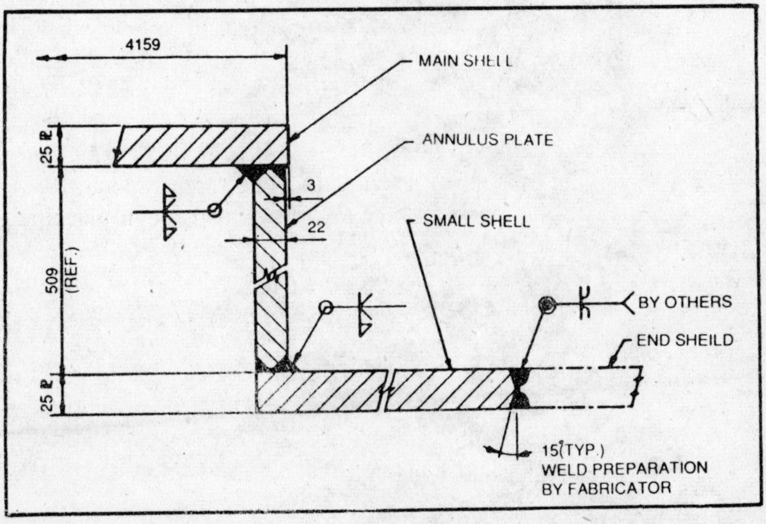

Fig. 14.3 a: TYP Annulus Plate to Main Shell Weld.

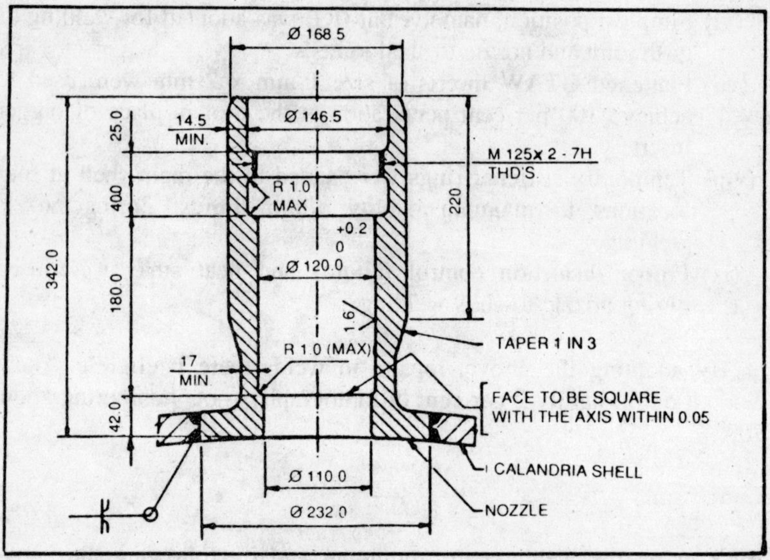

Fig. 14.3 b: TYP Machined Nozzle for NAPP.

SESSION IV

ALTERNATE SOURCES OF ENERGY IN J&K STATE

Chairman: Prof. M Ramakrishna Rao,
Indian Institute of Science, Bangalore

FIFTEEN

THE NEW AND RENEWABLE ENERGY THRUST OF THE J & K GOVERNMENT: A Factual Report

ARUN KUMAR

It is a well-known fact that the availability of energy in both commercial and non-commercial sectors is declining at a faster rate than can be imagined and therefore a nation like India does need the investment of scarce capital resources on a massive scale in, first of all, the commercial energy sector, in the areas of oil exploration and production, the exploration and exploitation of coal reserves, particularly those at great depths and of course the harnessing of hydel and nuclear energy. This investment is all the more necessary because, according to one calculation, one unit of investment in power leads to 10 of investment in downstream energy using productive sectors; which, in other words, implies that a reduction in such an investment would lead to tenfold idle (or lower) investment downstream with a larger actual reduction in output depending on capital-output ratio. There is thus no alternative to the massive investment in centralized electricity generating units for meeting the needs of the commercial sector.

However, at the same time, it will be no mean help if the non-commercial or the domestic sector becomes somewhat self-sufficient in its energy demands for cooking, heating, lighting and irrigation, all four sectors of which the first two are supposed to be extremely wasteful in J&K – so that this "surplus energy" also could then be made available for the commercial sector. Of course, this "selfish" reason is not the only logic dictating the J&K Government's efforts to find an appropriate answer to the energy crises in the domestic/rural sector because this alternative route to energy has by now become a necessity for our very existence. The outrageousness of this statement will somewhat lessen if we realize that many rural households (according to a Planning Commission study) are now wasting 6-24 man hours daily on collecting just their daily requirements of fuel and fodder. Children thus become economic assets in the short run which is all that matters and there goes all our family planning efforts out of the window. Women remain caught in the mire of their daily drudgery and that puts quite an effective end to all the slogan of women's liberation. Rural electrification here might provide an answer except that if this is to be done through large centralized plants over long transmission and distribution links; as in a hilly State like ours, it becomes a very expensive and, in the case of the not too infrequent natural calamities, a very unreliable proposition. But then if villagers are allowed to continue using fuelwood at the existing rate, the result will be deforestation with its attendant consequences of soil erosion, reduced productivity of land, increased siltation of waterways and increased flooding, reduction in rainfall, long-term climatic changes – in other words a total ecological disaster.

It is evident that to avoid all this a moratorium on tree felling specially in severely degraded forest areas and a massive programme of fuel-forestation and wood-plantation has to be undertaken on a war footing. Simultaneously, as also unanimously resolved by the U.N. Conference on New and Renewable Sources of Energy (NRSE) 1981, action is needed for reducing the pressure on fuelwood supplies by the development and popularization of such devices as can supply energy in a manner that can improve the standards of living and quality of life specially in our rural areas.

Such devices are quite well known and would for example include the adoption of improved fuelwood stoves, which not only reduce the quantity of wood required by 50 per cent but also improve the quality of life by emitting no smoke inside the kitchen. It has been calculated that adoption of one *Dhauladhar Chulha* for a year means the saving of four full-grown chir or trees which take almost 50 years to mature. The efficiency of these stoves can be further improved by introducing cooking-vessels of such conductivity which utilize heat in a more uniform manner and pressure cookers which cut down cooking time by a further 75 per cent. The Government can further popularize the adoption of solar cookers and bio-gas plants for providing another alternative to the widespread use of fuelwood.

Similarly the heating needs of our hilly and snow-bound areas can be reduced through the adoption of such housing designs which by using glass panes, black painted surfaces, thermal storage walls, etc., heat up the houses considerably. The requirement of hot water for the domestic sector can be similarly met by installing solar geysers of appropriate capacities. As for irrigation and lighting, solar, wind and bio-gas – all can be harnessed in series or parallel to come to our help.

The recommended philosophy for the Jammu and Kashmir Government is thus very clear, in which the Government is, of course, further guided by the objectives for the development of NRSE as embodied in the Sixth Five Year Plan of the Government of India which, because of their continued validity for the Seventh Five Year Plan, also deserve reproduction as below:

(a) to implement on a large scale programmes such as those of energy, forestry and bio-gas, where technology development has already reached a stage which permits field applications,
(b) to carry out field testing and demonstration on a country-wide basis of technologies which have the potential to become commercially viable in the next 5-7 years; and
(c) to intensify research and development of other technologies where the potential is likely to be available over a longer time horizon.

To this the Approach Paper to the Seventh Five Year Plan (1985-90) of the Planning Commission (pp. 16) adds and we agree that;

> Even though the current contribution of renewable energy sources to energy supply is insignificant and many of them are not commercially competitive at the moment, these technologies have to be developed now so that the transition in future will be smooth.

Past Experiences and Future Plans

In 1982, a small Division for Ecology and Environment was created in the State's Planning and Development Department to deal with NRSE which in 1985-86 transferred this entire sector to the Power Development Department for more effective implementation. Accordingly an annual plan of Rs. 30 lakhs for 1985-86 has been drawn up (Annexure I) and is now under implementation. The corresponding figure for the Seventh Five Year Plan from the State

Sector is Rs. 150 lakhs which with due cooperation from the Central Government is expected to exceed Rs. 800 lakhs. The sectorwise details are as follows:

Solar Energy

Solar electrification: No doubt, solar energy is plentiful, perennial, available free of cost and because of its high grade nature, convertible to most other forms of energy, be it heat, mechanical, chemical or electrical. And yet, going totally solar does not still promise us a utopian future free of load-shedding and climatic discomforts. The reason is naturally the high costs involved and this is regardless of the route taken to solar electrification – photovoltaic or thermodynamic. In fact, according to a study made by Shri T. K. Bhattacharya of Central Electronics Ltd., even if the solar PV array and related electronics are supplied free of cost in a user-owned, roof-top mounted, stand-alone system of 0.1-1 KW capacity, the cost of batteries only would make the cost of energy not less than Rs. 2.18 per Kwhr. Of course, if all the items are charged for the average energy cost becomes as high as Rs. 12.22 per Kwhr for the first 10 years with an over-all 20 years average of approximately Rs. 7.75 per Kwhr.

The economics involved appear to make solar electricity quite out of the reach of a small State like J&K – till we take the economics of rural electrification in hilly areas on account. A recent study by Shri C. R. Bhattacharya, Chief Engineer, Electrical, Agartala (Electrical India, October 31, 1984) points out that in the case of hilly areas where the average cost of rural electrification be Rs. 6.50 lakhs based on 10 km. length of new lines, population of the village concerned be less than 100 and load prospects be hardly 25 KVA in 15 years, a photovoltaic-lighting-cum-irrigation system would be quite cost effective when compared to either diesel electricity generation or electricity supplied through the grid. The comparative study is produced in Table 15.1

This conclusion is, in a way, also supported by the Advisory Board of Energy, Government of India, which in its "second Set of Recommendations (Feb. 1984)" in para 50 recommends the solar photovoltaics "use in remote locations like Ladakh and North-East, use in off-shore oil platforms, for cathodic protection of pipelines, railway signalling and meteorological stations". It is heartening to note, therefore, that in the case of our State, the Department of

Non-Conventional Energy Sources (DNES), Government of India, has already agreed to electrify a village each in Zanskar and Bani – both very far-flung areas of J&K State by supplying solar PV arrays with batteries of 12 KW and 1 KW respectively. The State Government on its part has already provided for Rs. 3 lacks for taking care of the expenses for civil works, maintenance, etc., in 1985-86. If this

TABLE 15.1: **Corporation in Rural Electrification and Photoveltion System**

Sl. No.	Item	Diesel	Rural Electri- fication line	Photovoltaic system
1.	Population	100/120	100/120	100/120
2.	Households (Nos)	20	20	20
3.	Cultivatible Land (HA)	15	15	15
4.	Pump load (KW) (Surface water presumed to be available for irrigation and with treatment can be used for drinking)	11	11	11
5.	Lighting load including St. Lights (KW)	3	3	3
6.	Demand day/night (KW)	11/2.5	11/2.5	11/2.5
7.	Installations	30 KVA	25 KVA	12 KVA
8.	Capital investment with cost of pumps & 0.5 km L.T. & H.T. lines (Rs. in lakh)	2.80	7.00	9.00
9.	Fixed charges (Rs. in lakhs) (Interest and Depreciation)	0.50 (18%)	0.81 (11.5%)	1.04 (11.5%)
10.	Operational cost (Rs. in lakhs)	0.28 (10%)	0.35 (5%)	0.18 (2%)
11.	Spares and Maintenance (Rs. in lakhs) (Energy need 7 hours for pumps for irrigation and drinking water for 7 months and 2 hours for drinking water and other purposes for remaining 5 months and 6 hours for lights).	0.28 (10%)	0.18 (2.5%)	0.9 (1%)
12.	Annual unit consumption	25000	25000	25000
13.	Annual cost (Rs. in lakhs).	1.06	1.30	1.31
14.	Per Unit Cost	Rs. 4.24p	Rs. 5.20p	Rs. 5.24
15.	Fuel/Energy Cost for (3) & (4)	Rs. 1.50p	Rs. 0.50p	–
16.	Total cost/unit	Rs. 5.74p	Rs. 5.70p	Rs. 5.24p

Note: Cost of diesel/lubricating oil is high due to transport bottlenecks. Rates and percentages are based on load conditions in Tripura and current practices, cost of irrigation pumps and batteries is included in photovoltaic system capital cost. Separate cost of pumps has been included in diesel power and rural electric lines. Consumption of energy to the level of 200-250 per capita would, in fact, reflect good standard of living and economic growth.)

projection succeeds, and if the Department of Non-Conventional Energy sources in view of the world-wide declining prices of solar PV cells agrees to aid us further, by 1990 J&K should be able to have a generating capacity of almost 150 KW of solar electricity. A provision of Rs.15 lakhs has, therefore, been made by the State Government for this purpose.

Solar Cookers: Cooking with sunlight is an attractive idea except that it involves the slightly uncomfortable prospect of cooking in bright sunlight, around noon, which further clashes with the normal mealtimes of our village-folk, which is either in the morning or around sunset. Then the existing box-type models are not able to generate high enough temperatures for frying or faster cooking. Design of a low-temperature pressure cooker, development of concentration type cookers which generate higher temperatures and allow a longer cooking time period of, say 8 a.m. to 4 p.m. the possibility of employing heat pipes to convey heat collected at a remote location to the kitchen and the storage of solar radiation in thermal, chemical or electrical devices, therefore, need to be investigated immediately especially with a view to making these affordable.

Till then, we seem to be stuck with the now familiar box-type, black painted, double glass covered solar cookers, whose sale has now started on a big scale from the Super Bazar, Jammu. After availing of a subsidy of Rs. 150 per solar cooker from the DNES, the cooker's sale price has been fixed at Rs.395. It has been proposed to further subsidize it by Rs. 50 from the State Government side, after which at Rs. 345 a piece, the cooker should prove to be popular in all parts of our State.

It is also proposed to equip the Angan Wadies/Bal Wadies of our State with community solar cookers at 100 per cent subsidy from DNES. Two such cookers are functional at the Bal Ashram, Kathua, since 1983-84.

Solar Water Heaters: The supply of low-grade heat (up to 140°C) through solar thermal devices is now an extremely viable proposition. From the point of view of energy conservation also, it has been calculated that the adoption of a 10,000-litres-per-day solar water heater would save 17,445 Kwhr of electricity per year. For 1985-86, the DNES sanctioned five community size projects for an outlay of Rs. 5 lakhs and 50 domestic size geysers for Rs.1.50 lakhs. Some of the projects worth mentioning are the 2,000 lpd at 70°C project for Government College for Women, Gandhinagar, Jammu, for Rs. 1.31 lakhs and the 3,000 lpd at 60°C project for Government G.M. Science College, Jammu, for Rs. 1.21 lakhs. It may be mentioned here that the DNES has been extremely encouraging in subsidizing such solar systems to the extent of 100 per cent in the case of Central Government, autonomous, educational, religious and charitable institutions, 75 per cent in the case of State Government institutions and cooperative societies and Municipal Corporations, 50 per cent in the case of Public Sector Corporations and so on.

It is hoped to complete all these projects in 1983-86 and by 1990 to have an installed capacity of a minimum of 100,000 litres per day of Solar Heated Water in Jammu and Kashmir. The sale of domestic Solar Geysers of 50-100 litre capacity is also proposed to start soon (at 50% subsidy) from the Super Bazar, Jammu, which has received the first lot of such geysers from the suppliers (Annexure II).

Solar Air Heating: One project for 1984-85 was sanctioned by the DNES at an outlay of Rs. 1.50 lakhs which would hopefully be installed in one of our public sector units for wool-drying purposes during the course of this year. Depending on the success of this project, by 1990 it is expected to complete 10 such projects.

Solar Kilns: The Forest Research Institute, Dehradun, has estimated that planting takes about three months to season in air, 10 to 12 days in steam heated kilns and about 18 to 24 days in a solar kiln. However, costwise while 25 mm planking in a solar kiln costs Rs. 161.30/m^3, in a steam kiln it comes to Rs.307.00/m^3. The initial installation cost also of a 7 m^3 solar kiln is about Rs. 65,000 which is 30 per cent of the initial installation cost of a steam-heated kiln of the same capacity, which comes to around Rs.1,55,000. The FRI, therefore, advises that,

> Besides complete drying to 12 per cent m.c the solar kiln has immense possibilities for safer accelerated predrying of green timber to about 25 per cent m.c. than normal installed steam-heated kilns to reduce overall time and cost of seasoning in a steam-heating kiln . . . the solar-heated kiln is the only low-cost appropriate technology presently available for accelerated drying of timber for cottage, and small-scale timber industries.

Considering the vast scope of this technology for a State like J&K which abounds in numerous timber-based industries, the DNES was kind enough to sanction two projects in 1984-85 for an outlay of Rs. 1.00 lakh. Hopefully these projects would be completed this year and by 1990, the State would have commissioned a minimum of 10 such projects. A provision of Rs. 2.50 lakhs has been made in the State's Seventh five Year Plan for this purpose.

Solar Dryers: In Jammu and Kashmir efforts have been made to popularise these devices in Ladakh for apricot drying and in Udhampur for chillies and *anardana* drying. The results obtained have been quite encouraging but the high costs involved (e.g. a 75-100 kg capacity solar drier was supplied at Rs. 16,400) have stood in the way of this device becoming attractive for individual users. Secondly, while these driers were of "Natural Convective Type" whose efficiency is supposed to be of the order of about 5-6 per cent for better consumer acceptance, we need to introduce "Forced Convective Type" dryers whose efficiencies can be of the order of 40-50 per cent. Finally, for attracting any subsidy from the DNES, it is necessary that these dryers be evaluated by the National Industrial Development Corporation Ltd, New Delhi.

All these problems are being sorted out and soon it is hoped to introduce community size solar dryers for drying grains, fruit, wool, leather, silk-cocoons, etc. An outlay of Rs. 2.50 lakhs has been kept for the purpose for the Seventh Plan period (Annexure-III).

Solar Stills: An attempt is being made to introduce these devices in Jammu and Kashmir for meeting the supply of distilled water for the vehicles of the State Motor Garages and the State Road Transport Corporation, for the hospitals in far-flung areas, and later, subject to feasibility, for drinking water purposes in Ladakh which abounds in salt-water lakes. For 1985-86, therefore, it is proposed to commission a plant of 5,000 litre/day capacity in Jammu for the State Motor Garages at an outlay of approximately Rs. 3 lakhs. Other projects will be taken up afterwards.

Solar Architecture: Adoption of Solar Thermal Storage Wall (trombe walls) and simple "glass" houses have been quite popular in the Ladakh region for quite some time. In the State sector, however, a project was sanctioned in 1982 in favour of the Regional Engineering

College, Srinagar, for Rs. 4.88 lakhs of which the DNES contribution was Rs. 3.44 lakhs. Some more projects have been sanctioned by the DNES in Ladakh at an outlay of over Rs. 50 lakhs. The successful commissioning of all these projects is being eagerly awaited. The DNES is further understood to have sanctioned two more such projects for the ITBP and the Army in Ladakh for an outlay of over Rs. 50 lakhs on 100 per cent subsidy basis. We are eagerly awaiting the commissioning of these projects and depending on their success plan to spend up to Rs. 5 lakhs in the State sector to popularize this idea in the Seventh Five Year Plan.

Wind Energy

Water Lifting Pumps: Under a National Project for the demonstration of these simple devices, 25 wind pumps were allotted to Jammu and Kashmir (on full payment for five only) at the rate of Rs. 9,500 per wind pump) and four separately free of cost to Kashmir Government Polytechnic, Srinagar, ostensibly for the purposes of study and further fabrication. These pumps are designed to operate at a cut in velocity of 2.5 m/sec and give a discharge of almost 20,000 litre/hr at the cutout velocity of 10m/sec. In all, 16 wind pumps were received from the manufacturers, but only one could be installed in Tikri, District Udhampur, for irrigating a forest nursery. The reasons for this low achievement of targets range from the unfamiliarity of our engineers with these devices to the bad/indifferent results obtained from the one wind pump installed so far. Anyway efforts are on to sort out these problems and, hopefully, all the 16 wind pumps will be installed to our satisfaction during the course of this (1985) year. We have also contacted the Indian Institute of Sciences, Bangalore, (Department of Mechanical Engineering) for their advice in this regard.

Aero-Generators: No reliable wind velocity data exists for our State specially for the sites which would be suitable for the installation of aero-generations. Therefore, a list of nine sites which include Choglansar (Leh), Nagri, Kaskoor (both in Doda District), Nathatop and Guntop (Udhampur District), Zozila Pass (Kargil), Rajdingini Pass (Baramulla), Nastacheen Pass (Kupwara) and Shey/Thiksy/Igo along Indus River (Dist. Leh) has already been sent to the DNES for the installation of suitable anomographs. Apart from these, the DNES has already agreed to carry out the wind velocity survey of Haftal plains for the installation of a 100-KW aero-generator and have asked M/s J. K. Synthetics, Kanpur, to do the needful. In the meanwhile, at Karthgali, Dist. Udhampur, after a casual survey by the engineers of the Wind Energy Research & Development Centre, Allahabad, it was decided to install four 3-KW each aero-generators of which two are in advance stages of installation.

Hopefully these preliminary attempts will pave way for the commissioning of a full-fledged wind farm in our State in megawatt ranges at some suitable site. The Power Development Department has, therefore, made a provision of Rs. 15 lakhs on this account for the Seventh Five Year Plan.

BIO-GAS

Family size Units: Family size bio-gas units are being popularized in our State by the Agriculture Production Department and the KVIC, who by now have a trained cadre of masons and engineers. The cost of these units is fully met through DNES subsidies and bank loans. There have been certain problems in the operation of these plants during winter in Kashmir which needs looking into by scientists and for which a provision of Rs. 10 lakhs has been made for the 1985-90 period.

From this year onwards, however, stress is going to be given to the commissioning of large-sized community/institutional bio-gas project, which also attract 75-100 per cent subsidy from the Central Government. A total of Rs. 15 lakhs has been proposed for the State share for the 1985-90 period in which it is hoped to cover almost all major sheep/animal husbandry farms of the State Government.

Night Soil Based Plants: The Government has already commissioned the services of M/s Sulabh International, the famous voluntary agency from Bihar, whose work in electrifying a big stretch of road in Patna through sewage gas has earned widespread appreciation. Already five project reports for different parts of Srinagar have been prepared and although these would be primarily sanitation improvement projects the Power Development Department would like to study the feasibility of gas/electricity generation wherever possible. The DNES has, therefore, been already contacted and we, on our part, have made provision of Rs. 5 lakhs for the 1985-90 period for such projects.

Integrated Sanitation Projects: These projects are supposed to deal with the sanitation problems of entire cities and therefore are to be handed by the Urban Environmental Engineering Department. However, the Power Development Department intended to involve itself with such projects just to ensure that (a) all possibilities of bio-gas generation from such projects on the lines of, for example, Okhla Project in New Delhi, which is now supplying cooking gas to more than 2,000 families, be explored, and (b) electricity be generated from such gas wherever it proves to be cost effective.

A sum of Rs. 5 lakhs has been kept for this purpose in the Seventh Five Year Plan.

Weed Gasification Project: This is also an offshoot of the lake beautification projects of the Urban Environmental Engineering Department who are reported to be extracting up to three tonnes/day of weed from the Dal Lake alone. If gasified and supplied to nearby hotels for cooking purposes, a saving of a minimum of 35,000 litres of kerosene oil is calculated to be effected every year.

The Power Development Department has, therefore, initiated steps to identify suitable consultant for the design and execution of such a project and has earmarked Rs. 10 lakhs for this purpose for the 1985-90 period.

Bio-Mass

The concept of planting fast-growing species to meet the fuel and fodder needs of our populace is already well established in Social Forestry circles. The Power Development Department would, therefore, like to involve itself in this field only to ensure that the fuel needs of rural areas are well taken care of by the plantation programmes of the Forest and Rural Development Department so that the cooking load on electricity is brought down. At a later stage it is proposed to go in for "Captive" lots of fast-growing energy plantations for wood-gasifier non-generators. Such generators can also run on saw dust, rice husk, etc., which are available in plenty in the State. In case such waste material has to be compacted (with or without binders) before they can be fed to gasifiers the Power Development Department plans to set up such briquetting plants also on its own, if necessary.

The technology for the above concepts is now available in our country and some consultants have already been identified. A sum of Rs. 25.50 lakh is, therefore, kept for experimentation in this very promising field.

Another measure to reduce energy consumption in this sector is the popularization of "Smokeless Chulhas" of which the fixed types are subsidized 100 per cent by the DNES and the portable ones to the extent of 75 per cent. Both the varieties have been introduced in the Jammu region of our State and have proved to be extremely popular. Simultaneously, the DNES has also launched a programme for training people, especially womenfolk, in the art of erecting such *Chulhas*, and already there is a cadre of over 2,000 persons in the State. During the course of the Seventh Five Year Plan now, attempts shall be made to introduce these *chulhas* on a war footing in almost every corner of the J&K. It is also proposed to introduce this concept for the owners of restaurants, bakeries, dyers, washermen, etc., who also are great guzzlers of coal/wood and are great polluters at the same time.

Geo-Thermal Energy

It is a matter of pride (and good luck) for our State that after 10 years of continuous testing, the Geological Survey of India has rated the hot water springs of Puge Valley, Ladakh, to be the most promising site for harnessing geo-thermal energy in India. Already boring to a level of 224,300 metres has been conducted and a temperature of 123.4°C obtained. During the course of this 1985 year if the ONGC also helps it is hoped to drill up to 500 metres and hit a temperature of 200°C.

The Central Electricity Authority has, therefore, already decided to set up a pilot 1-MW project at Puge and has probably requested the National Aeronautical Laboratory, Bangalore, to do the needful. The latter, it may be mentioned, has already set up a 5-KW binary cycle plant at Mapikaran in Himachal Pradesh. The Power Development Department has, therefore, made a provision of Rs.10 lakhs on this account for the 1985-90 period.

The hot water of Puge has also been in the meanwhile successfully used for heating some Army barracks nearby by the RRL Jammu which throws up many exciting possibilities of producing vegetables, fruits and medicinal plants in green houses and running fishery, poultry, sheep breeding, wool/food processing units etc. The District authorities of Leh have therefore been already contacted in this regard.

Non-conventional Refrigeration and Air-Conditioning

An underground refrigeration project for storing vaccines at the desired temperatures has been sanctioned for 1985-86 for Leh with the active collaboration of UNICEF and the Tata Energy Research Institute at an outlay of Rs. 2 lakhs. This project is expected to be commissioned during the course of the present year (1985).

Depending on the success of this project more such projects will be launched in different parts of our State for storing vegetables, fruits, medicines, etc. The feasibility of harnessing such alternative energy sources as solar, wood/bio-gas, cold river water (such as that of the Chenab), etc; can also be explored for this purpose. A sum of Rs. 5 lakhs has therefore been proposed for the Seventh Five Year Plan under this head.

Non-conventional Water Lifting Devices

Apart from wind pumps this sector envisages the widespread propagation of solar/wood-gas/bio-gas pumps and hydrams for drinking water and irrigation purposes. Hydrams, which are fuelless pumps, and which lift water up to 20 litres the fall of water by utilizing the power of the waterflow itself, have been successfully introduced in Khalsi (Ladakh) and Shalimar and Pattan (Kashmir). Such waterlifting devices are eligible for further subsidies from the Ministry of Irrigation, Government of India, and refinance facilities from NABARD.

A sum of Rs.5 lakhs for the 1985-90 period has therefore been proposed from the Power Development Department outlay also for encouraging the use of these devices.

Integrated Rural Energy Programme

Under this newly launched programme of the DNES, entire villages are taken up for equipping them with every non-conventional energy device available through the aegis of the DNES completely free of cost. It has been, therefore, proposed to have one such demonstration village in each district in the Seventh Five Year Plan.

Meteorological Stations

The importance of collecting data regarding solar intensity, wind velocity, rain, snowfall, etc., can be hardly overemphasized in the context of the proper harnessing of NRSE. Accordingly, the matter has been taken up with the Central Government so that such stations could be set up by the Indian Meteorological Department. The Power Development Department, on its part, however, has kept a token provision of Rs.2.50 lakhs for the 1985-90 period for sharing expenses on survey, maintenance, collection of data etc., if it be required to do so.

The Power Development Department has further entered into an agreement with the National Remote Sensing Agency, Hyderabad, for sharing data on snow-mapping of our State and it is in the process of getting some of its staff trained in this very specialized subject. The State Government is also considering a proposal for the setting up of a full-fledged Centre for Remote Sensing in J&K so that the other advantages of remote sensing in the field of surveys of crop/forest conditions, geological mapping, soil mapping, etc., could also be fully availed of.

Waste Heat Recovery and other Energy Conservation Projects

It is a well-known fact that a thermally driven engine producing electricity has a typical conversion efficiency of not more than 33 per cent. In places like Ladakh where atmosphere is even rarer, this conversion efficiency is usually much less. The remainder energy from the fuel is naturally rejected by the engines as waste heat. Theoretically speaking, therefore, a diesel generator providing 80 KVA electricity in Leh will be producing at least 50 KW of waste heat, which is sufficient to heat up to 9,800 litres of water in five hours.

Since Leh and Kargil Districts alone have some 50 diesel generators of derated capacities exceeding 2,800 KVA it should be possible to tap an enormous quantity of hot water, which can be used if for nothing else than for just bathing purposes. Negotiations are, therefore, going on with UNICEF and the Tata Energy Research Institute for sharing the cost of first such project which can be expected to be commissioned within the course of this year (1985). Elsewhere, the hot water so produced can be used for refrigeration purposes.

Other energy conservation measures sought to be examined in this context are (a) improvement of efficiency of all kinds of pumps used in our State in accordance with the recommendations of the Advisory Board of Energy, Government of India, (b) off-peak or time-of-the-day tariff to discourage peak hour loads, (c) provision of energy audit services to industrial bulk consumers, etc.

Conclusion

Admittedly the above list of items constitutes quite a tall order whose successful execution might not be possible to the desired extent by the existing machinery of the Power Development Department. Many other States have therefore already formed separate corporations or agencies to deal with this new subject in exclusive manner. There may be some merit in adopting this course of action for J&K State also and, therefore, a full-fledged proposal has been sent to the J&K Government for consideration. Hopefully with the setting up of a new body for this purpose the programme for the harnessing of new and renewable sources of energy will get the much needed fillip here.

In any case, this is a field which needs great collaborative efforts from all concerned be they academicians, scientists, private manufacturers, social workers or media men. It is hoped that the present exercise will go to contribute to a little extent in that worthy direction.

ANNEXURE I

Seventh Five Year Plan 1985-90 and Annual Plan 1985-86
(State J&K)

(Rupees in Lakhs)

Sl. No.	Name of Scheme	Allocation Centre 1984-85	Actual Expenditure 1984-85	Seventh Plan (1985-90)		Annual Plan 1985-86	
				Outlay Stage	Outlay Centre (Expected)	Outlay State	Outlay Centre (Expected)
1	2	3	4	5	6	7	8
1.	Solar Electrification	nil	0.023	15.00	100.00	3.00	13.00
2.	Solar Cookers	6.00	nil	2.50	5.00	0.50	1.00
3.	Solar Air Heating	1.50	nil	2.50	10.00	0.50	1.50
4.	Solar Water Heating	6.50	nil	2.50	10.00	0.50	1.00
5.	Solar Timber Kilns	1.00	nil	5.00	5.00	1.00	1.00
6.	Solar Driers	0.50	nil	2.50	5.00	0.50	1.00
7.	Solar Architecture	nil	nil	5.00	10.00	1.00	1.00
8.	Solar Stills	nil	nil	3.75	5.00	0.75	1.00
9.	Wind Pumps	nil	nil	2.50	5.00	0.50	nil
10.	Aero-generators	nil	nil	15.00	100.00	3.00	10.00

Contd.

1	2	3	4	5	6	7	8
11.	Bio-gas (Community) Inst.	nil	nil	10.00	100.00	2.00	10.00
12.	Sewage Gas	nil	nil	5.00	100.00	1.00	10.00
13.	Integrated Sanitation Projects	nil	nil	5.00	10.00	1.00	1.00
14.	Weed Gasification	nil	nil	10.00	40.00	2.00	8.00
15.	Wood Gasification	nil	nil	17.50	50.00	3.50	5.00
16.	Briquetting/Pelletization	nil	nil	7.50	15.00	1.50	3.00
17.	Energy Plantations	nil	nil	0.50	2.50	0.10	0.50
18.	Smokeless Chulhas	5.35	1.35	2.50	10.00	0.50	2.00
19.	Geothermal Energy	nil	nil	10.00	100.00	2.00	10.00
20.	NRSE-Refrigeration & Air-Conditioning	nil	nil	5.00	10.00	1.00	2.00
21.	NRSE Water-lifting devices	nil	nil	5.00	10.00	1.00	2.00
22.	Integrated Rural Energy Programme	nil	nil	3.25	10.00	0.65	2.00
23.	Meteorological Stations	nil	nil	2.50	10.00	0.50	2.00
24.	Waste Heat Recovery & Other Energy Conservation Measures	nil	nil	5.00	nil	1.00	nil
25.	Misc. Administrative Expenses	nil	nil	5.00	nil	1.00	nil
	Total	20.85	1.373	150.00	712.00	30.00	88.00

ANNEXURE II

Payback Period Analysis for 250 LPD at 60°C Capacity Domestic Solar Water Heating Systems

		UTILIZATION FACTOR					
		30%	60%	100%	30%	60%	100%
1.	Collector Area sq. m.	5	5	5	5	5	5
2.	Average solar insolation on inclined surface MJ/d/m^2	21	21	21	21	21	21
3.	System efficiency,%	50	50	50	50	50	50
4.	No of days of effective operation in a year	100	200	300	100	200	300
5.	Energy saving by the solar system in a year: in MJ	5250	10500	15750	5250	10500	15750
	in KWH	1458	2917	4375	1458	2917	4375
6.	Replacement fuel considered and Unit rate	Electy.	Electy.	Electy.	Electy.	Electy.	Electy.
	per KHW	Rs. 0.70	Rs. 0.70	Rs. 0.70	Rs. 1.00	Rs. 1.00	Rs. 1.00
7.	Reduction in Elect. bill	Rs. 1021	Rs. 2042	Rs. 3063	Rs. 1458	Rs. 2917	Rs. 4375
8.	Capital cost of system.	Rs. 12000	Rs. 12000	Rs. 12000	Rs. 12000	Rs. 12000	Rs. 12000
9.	Payback period:						
	(i) without subsidy	—	12 yrs	6 yrs	—	7 yrs	4 yrs
	(ii) with subsidy (50%)	12 yrs	4 yrs	3 yrs	7 yrs	3 yrs	2 yrs.
10.	Payback period with 10% annual escalation in Electricity cost:						
	(i) without subsidy	—	8 yrs	5 yrs	11 yrs	5 yrs	4 yrs
	(ii) without subsidy (50%)	8 yrs	4 yrs	3 yrs	5 yrs	3 yrs	2 yrs

Courtesy: Gujarat Energy Development Agency.

ANNEXURE III

Payback Period Analysis for Industrial Solar Water and Air Heating Systems

Replacement Fuel	Hot Water System			Hot air System		
	Coal	Fuel Oil	Electricity	Coal	Fuel Oil	Electricity
1	2	3	4	5	6	7
1. System capacity	5000 lpd at 60°C	5000 lpd at 60°C	5000 lpd at 60°C	4000 m^3/ hr at 80°C	4000 m^3/ hr at 80°C	4000 m^3/ hr at 80°C
2. Collector area, sq. m.	100	100	100	120	120	120
3. Average solar insolation on inclined surface, MJ/d/m	21	21	21	21	21	21
4. System efficiency, %	50	50	50	50	50	50
5. No of days of effective operation in a year	300	300	300	300	300	300
6. Energy saving by the solar system in a year:						
in MJ	315000	315000	315000	378000	378000	378000
in KWH	87500	87500	87500	105000	105000	105000

Contd.

1	2	3	4	5	6	7
7. Unit rate per KWH of replacement fuel	Rs. 0.24	Rs. 0.34	Rs. 0.40	Rs. 0.24	Rs. 0.34	Rs. 0.40
8. Reduction in fuel bill	Rs. 21000	Rs. 29750	Rs. 35000	Rs. 25200	Rs. 35700	Rs. 42000
9. Income tax on savings due to solar system @55% + 2.5% surcharge thereon, i.e.Total @56.375%	Rs. 11839	Rs. 16772	Rs. 19731	Rs. 14207	Rs. 20126	Rs. 23678
10. Net savings due to Solar Systems	Rs. 9161	Rs. 12978	Rs. 15269	Rs. 10993	Rs. 15574	Rs. 18322
11. Capital cost of the system	Rs. 175000	Rs. 175000	Rs. 175000	Rs. 200000	Rs. 200000	Rs. 200000
12. Payback period:						
(i) One shift op.	10 yrs.	7 yrs.	6 yrs.	9 yrs.	7 yrs.	6 yrs.
(ii)Three " "	9 yrs.	6 yrs.	5 yrs.	8 yrs.	6 yrs.	5 yrs.
13. Payback period with escalation in fuel cost:						
(i) One shift Op.	7 yrs.	6 yrs.	5 yrs.	7 yrs.	6 yrs.	5 yrs.
(ii)three " "	7 yrs.	5 yrs.	4 yrs.	6 yrs.	5 yrs.	4 yrs.

SIXTEEN

ENERGY FROM BIO-MASS

M.M. BOKADIA

Biomass

The term biomass, now universally accepted in energy writings, refers to any plant mass harvestable and wild for conversion to fuel, as well as to animal and human wastes convertible to solid or gaseous fuels. As these wastes are overwhelmingly nothing but undigested residues of plants, the term phytomass would be preferable, with biomass reserved for both of the living masses making up the biosphere, the mass of autotrophs (phytomass) and the mass of heterotrophs (zoomass, including humanity for which a separate term anthropomass is also used).

In any case, most of the planet's biomass is phytomass. In ecological definition biomass is the dry matter of living organisms present at a given time per unit of the earth's surface. It is present to the extent of 45 kg/sq. metre for tropical rain forest or virtually zero for sand/ice deserts.

Anaerobic Fermentation

A vast amount of experience has already been gained in treating the biomass anaerobically and producing biogas. The treatment provides both fuel and fertilizer and cannot help but improve environmental hygiene and reduce objectionable odours and pollution. However, biogas production involves a host of environmental, technical, economical, cultural and social factors which combine to make the task of spreading anaerobic digestion a rather difficult one. In recent years specialized biochemical and engineering literature on the processes, products, requirements and efficiencies of anaerobic fermentation has been greatly enriched particularly during the 1970s, due to extensive researches both in the United States and China.

Methanogenic Fermentation

Anaerobic decomposition proceeds in three phases:

First, bacterial hydrolysis (an enzymatic breakdown of polymeric carbohydrates, proteins and lipids) followed by an acidogenic phase (fermentation of monomers or dimers of the carbohydrates and simple sugars in the food stock to acetic acid, propionic acid, CO_2 and H_2) and finally by methanation (in which methanogenic bacterial utilize NH_3 as a source of N and release a mixture of CH_4, CO_2 and traces of several other gases. The whole process is carried on simultaneously in an anaerobic digestor and is thus fairly complex, but fortunately it is self-regulatory to a large extent.

Waste can be fermented in batches or in a continuous process, the latter being overwhelmingly the method of choice for an efficient operation. Raw materials fed into a digester (airtight) above ground or in-ground metallic, masonry or plastic container can be led directly from stables, latrines or they can be stored before the loading, pretreated and reduced to a uniform size by chopping or shredding. Dilution is almost always necessary so that the slurries contain only 8-12 per cent solids, otherwise the material is difficult to handle and fermentation is impeded.

A digester is usually started by seeding it with an appropriate mass of both acidogenic and methanogenic bacteria and the subsequent loading rate (weight of volatile solids per unit of digester volume per day) varies with the waste material and its rate of decomposition. Mixing the digester's content promotes contact between microorganisms and the substrate and is thus essential for good conversion efficiency. If left undisturbed, the loaded material separates into layers with scum floating on the top of the supernatant above the active and stabilized solids at the bottom. The rate of methanogenesis slows down and the evolving biogas has difficulty in escaping through the scum lid. Thus scum is one of the major dister problems and must be avoided at all costs (John L. Fry 1974).

Fermentable Material

Animal waste, human waste, crop residues, wild grasses, household garbage, etc. Compost all the plant material: straw, stocks, grasses and weeds are chopped, piled up in layers sprinkled with lime and manure and sealed with clay for a week to 10 days in summer, much longer in winter. Partial decomposition of the plant matter lowers its C:N ratio.

Maintaining a near optimum carbon: nitrogen ratio is another critical concern. If C:N ratio be too high the process will be limited by nitrogen availability and the bacteria will reproduce only slowly, should it be too low high ammonia concentrations can become toxic to the bacteria. The optimum ratio by weight is close to 30 and the substances which come closest to it are animal manures (15-20), raw household garbage (25), grasses and weeds (15-20). The most abundantly available biomass materials wood (saw dust) and untreated crop residues have very high C:N ratios (excess of 50). Animal manures are mixtures of wastes with lower than optimum C:N ratios (faeces and urine) and materials having ratios higher than optimum (bedding, straw or sawdust) and are thus ideal fermenting materials. In contrast biogas production from crop residues alone cannot proceed at all well. Maintenance of adequate nutritional content for the growing bacteria thus puts clear limits on the admixture of crop residues, leaves or food processing residues to anaerobic digesters fermentating animal wastes. (10 per cent human excreta, 40 per cent animal manure with crop stalks or grass and the rest water).

Biodegradability

Lignin remains practically unaffected by anaerobic fermentation and cellulose is broken down only very slowly. Poultry wastes have the highest fraction of volatile solids convertible by anaerobic fermentation.

Undigested straw in animal excreta is cellulose and hemi cellulose. The residues cellulose and hemicellulose are in theory completely hydrolysable to simple sugars which in turn can be completely converted by methanogenic bacteria to gas.

Temperature

By far the most decisive environmental limitation in the process of anaerobic fermentation is the temperature. Methanogenesis proceeds best in the range of 20-60°C with the optimum for mesophilic bacteria at around 35°C and for thermophilic bacteria 55°C. The thermophilic fermentation cannot be sustained without an outside source of energy and various designs ranging from soil immersion heaters to solar devices (Green House Technique) have been tried. Heating the digester's contents will thus decrease the net energy available from the fermentation. But even in colder climates the operation may be made self-sustaining near the optimum temperature by diverting a part of the generated biogas to heat the influents and the digester's contents. For example a well-insulated pilot plant digester fermentation swine wastes in Manitoba and using the produced biogas for heating had a net energy recovery of nearly 40 to 50 per cent during winter (January in southern Manitoba averages around -10°C) and 55 to 100 in summer. In northern New York State fermenting the dairy manure in the winter energy recovery from 100 cows was nearly 52 per cent (the winter average is -5°C) (lower net energy return during winter).

Special measures are recommended for winter generation: Insulation by covering the pits with earth, straw or grass, more frequent and more voluminous lending and removal with feed stock containing relatively more human or animal excreta (every day if possible). Very frequent stirring and more assiduous maintenance of pH and abundant reseeding before the winter is recommended. The winter rates are at the best half of the summer performance. Underground installation of the digester has also been recommended.

Optimum pH

The optimum pH in municipal sewage digestion is very near neutral 6.6-7.6. But with animal manures the pH can be as high as 8. In extensive experiments at the University of Manitoba, Lapp and his colleagues (1975) found that alkalinities up to pH 8.5 were tolerated after acclimatization of methanogenic bacteria without any apparent reduction of methane production. The pH is maintained slightly alkaline and acidity is corrected by addition of lime or ash.

Thus the conversion efficiencies of anaerobic fermentation depends upon the kind and qualities of food stock, temperature of the influent and the digester's contents, loading rates, liquidity of the wastes, detention time, the period lapsed from seeding the digester, rates of gas generation, pH of the fermenting mixture and frequency of mixing the load mixture.

Some idea regarding the generation is given in Tables 16.1 and 16.2.

TABLE 16.1: **Biogas Production**

Fermenting wastes	m^3/kg of dry/solid	Temperature (°C)
Cattle manure	0.23–0.50	11.1–31.1
Night soil	0.38	20.0–26.2
Straw	0.2	20.0–25.0

TABLE 16.2: **Indian Biogas Generation**

	Prasad and Prasad Reddy	Makhijani and Poore	Parikh and Parikh
Dry Dung (kg/cow per day)	3.18	2.96	2.5
Collection efficiency	75	70	80
Biogas yield (m^3/kg dry dung)	0.187	0.382	0.180
Daily biogas production in a village with 500 cows (m^3)	111	198	90
Burned in traditional manner the efficiency is only 10 per cent			

Thus each of 300 cattle in a typical village would produce 3 kg of dry dung daily, 75 per cent of this would be collected and 0.31 m^3 of biogas containing 23 MJ/m^3 would be generated per kg of dry dung, 46 per cent of the produced biogas would go for cooking for 200 families and the rest for power generation.

Chinese Experience

The Chinese in 1958 inspired the peasants to build biogas digesters. In 1974 Sichaun had from 30,000 units to 120,000 units. By the summer of 1976 1.7 million Sicharanese peasants were using biogas for cooking and lighting. In 1977 the province had 4.3 million units providing fuel to 20 million villagers. Nation-wide total reached 7 million in August 1978. But in India there were only 15,000 units by 1984. The ratio per 10,000 rural families is 400 units for China, 36 units for Republic of Korea, 24 units for Taiwan and 1 unit for India.

Physical Properties

Biogas is normally a colourless, flammable gas composed of 60 to 70 per cent methane, 30 to 40 per cent CO_2, a fraction of one per cent of H_2S and traces of H_2, NH_3 and nitrogen oxides. Predominantly cellulsic feed stocks will generate relatively more CO_2 while higher proteins and lipid contents in the waste is conducive to higher methane share. H_2S is a problem for internal combustion because it will produce SO_2 and SO_3 which corrode the metal. The energy content of biogas ranges between 22 and 26 MJ/m^3 with 60-70 per cent methane. (43.1 MJ for fuel nil and 47.1 MJ for gasoline).

The Sludge

The fermentation process removes C, H and O preserving everything else in a sludge which may be used not only as a fertilizer but also as a feed stock for protein recovery or even for direct feeding. The sludge would give virtually all the nutritional benefits of unfermented manure without malodorous.

The total volume of effluent is first slightly less (2-5%) than the input and the organic fraction of the sludge (30 per cent of the original feed stock) is a mixture of lignin, lignin procted material, undigested cellulose and lipids, bacterial cellular substances and volatile organic fatty acids. As for the conserved nutrients 81.1 per cent of the loaded N is recovered in digested slurry after evaporating to dryness.

The effects of the applied sludge will be the enrichment of the soil by essential plant nutrients and improvement of the soils and physical properties, increased aeration, brightened moisture holding, water in filtration and cation exchange capacity. For applying the sludge the applier has to be equipped to spread liquid manure. What may cause problems is the temporary storage of sludge which cannot be applied immediately after removal. Improperly stored sludge may be a source of local water pollution.

The principles have been well known for a long time and municipal sewage digestion is a mature science but anaerobic fermentation on the farm is the vogue of 1970s. Hence the experience is far from extensive.

In the United States anaerobic fermentation has received considerable research attention.

Biogas can be used for household lighting, and water heating and in internal combustion engines, where in comparison with other common fuel it has a low heat content per volume, high octane number and low speed of combustion. Cooking needs 0.3-0.4 m^3/day; gas lighting needs only 0.07-0.08 m^3/h per mantle.

Advantages

(i) Saving fuels, (ii) Requiring less labour, (iii) Saving fuelwood and grasses, (iv) Saving straw and other crop residues for animal fodder and bedding, (v) Reducing expenditure of fuel, (vi) Lighting, (vii) Elimination of many insect pests and diseases and markedly improving hygienic conditions of rural areas, (viii) Protecting forests and timber, (ix) Local generation of electricity, (x) Production of an excellent organic fertilizer. In comparison with *Aerobic* fermentation the *Anaerobic* process results in superior products: it increases ammonia content by 120 per cent and the amount of quick-acting phosphorous by 150 per cent. A family digestor (6+6 cattle) can yield 100 kg of fertilizer twice a year. The sludge contains about 650 ppm of N, 40 ppm of P and 9,400 ppm of K. The sludge is used directly as field manure for basal or top dressing and also for making humic acid fertilizer. When it is not immediately required it is stored in heap plastered with clay. The effluent is regularly removed with the addition of new material. This is either applied directly to the land with irrigation water or stored in tanks for later use. However, its nutrient content is significantly lower than that of sludge.

It helps in saving coal, wood and crop residues and protecting environment. Burning all the residues as fuel is loss of organic matter and crop nutrients. Wood for fuel is a large fraction of the total forest biomass harvest and a major contribution to deforestation and erosion.

Time, labour and money saving could be impressive. In many villages people walk up to 5 km away to cut fuelwood, or to buy dried grass, coal or kerosene. Cutting transporting and handling of fuelwood would save man-days and money that is used in purchase of commercial fuels or firewood.

Moreover cooking with biogas is both faster and easier than with straw or wood and lamps lighted with gas are brighter.

Sanitation and Environment

Collection of all types of waste products will lead to better sanitation and a clearer atmosphere and would control spread of many diseases. In China many fatal diseases have been controlled with the setting up of biogas digesters.

SEVENTEEN

HYBRID MICRO HYDEL
Alternate Energy Source for Jammu and Kashmir

O.D. THAPAR

INTRODUCTION

Development, welfare and prosperity of the rural areas where bulk of our population lives is dependent upon the availability of adequate amount of energy for irrigated agriculture, rural industries and domestic needs. A large potential of energy remains untapped in flowing streams, irrigation work and river slopes and is available near rural consumption centres. This energy can be converted into electricity.

Jammu and Kashmir State on the north-western tip of India is almost entirely mountainous broken by valleys. It has a population of 59.54 lakhs (1981 Census). More than 80 per cent of the population depends on agriculture. Paddy, wheat and maize are the major crops. Horticulture and forestry are also important resources of the State. Hydro-power from snowfed streams and rivers is the most important energy source.

Large hydro-electric plants are highly capital-intensive, remotely located, have long gestation periods and impoundments are a potentional environment hazard. Small hydro can be extensively developed to exploit the available resources especially to meet rural energy needs.

Further, electrical energy being the highest form of energy can be used to upgrade other forms of energy in hybrid systems. Accordingly surplus electrical energy (during periods of low electrical demand) can be hybridized with solar and biomass which are other renewable energy resources available in rural areas.

A hybrid system integrating these sources of energy is under development for use in the rural areas of different States of India by the Alternate Hydro Energy Centre (AHEC). Hybrid system development has a strong potential for increasing the income and well-being of rural areas as outlined below:

(i) Increase agricultural output due to availability of energy (for agricultural operation) and fertilizers,
(ii) Timely availability of critical inputs to villagers in terms of energy and commodities,
(iii) Generate more employment in villages,
(iv) Provide alternate domestic fuel (save trees), utilize agricultural and forest waste (increase rural income) and hybrid system centre provide nucleus for integrated rural development, and
(v) Improve life-styles and tend to arrest population migration to towns.

This paper discusses the equipment and system option available for setting up an economically viable decentralized micro/mini hybrid hydel system in Jammu and Kashmir State to meet rural energy needs. The discussions are based upon demonstration projects being set up by AHEC.

Resource Assessment

Present hydro-power potential of Jammu and Kashmir is estimated at 5,000 MW and about 5 per cent is so far exploited. It is considered that at least 10 per cent, i.e. 500 MW can be developed as small hydro. This appears conservative.

Biomass energy is available from agro-wastes and forest residues. The amount of agricultural wastes would depend on the extent of agricultural production. The estimate of available agro-wastes based upon agricultural production in the State is given in Table 17.1.[2]

It is considered that about 20 to 30 per cent of agro-wastes can be made available for hybridization.

Data regarding forest residue comprising timber saw dust, wood waste, fruit tree pruning, etc., is not readily available.

Solar, wind, geothermal and animal dung energy potential is also available in the State and its potential needs assessment.

TABLE 17.1: **Grain Production and Agro-waste Availability**

Crop	Area under cultivation ('000 hectares)	Grain Production ('000 tonnes)	Aggro-waste ('000 tonnes)
Rice	237.1	426.78	896.24
Maize	267.0	359.38	539.07
Wheat	190.4	194.40	320.76
Bajra	15.2	8.19	13.92
Small millets	21.6	10.80	13.50
Pulses	49.0	29.60	47.36
Total	780.3	1029.15	1830.85

Small Hydro Power Technology

Small Hydro Power (SHP) Technology may be discussed under following sub-heads:

(a) Main Generating Equipment (turbines, generations, etc.)
(b) Controls (speed governing, voltage control, etc.)
(c) Civil Works
(d) Utilization

Cost of mini hydro power is highly site specific. Present technological improvements are directed towards reduction of cost by adopting standardized turbine and generator sets, simplified controls, simple civil engineering and utilization of local manpower and materials. The main problems in the development of small hydro plants are shown in Fig. 17.1[1] and are summarized below:

1. High capital costs.
2. Equipment availability especially in ultra low head (below 3 metres) range.
3. High cost of controls and management.
4. Low utilization (load) factor.

Typical cost distribution for low head and high head development are given in Fig. 17.2.

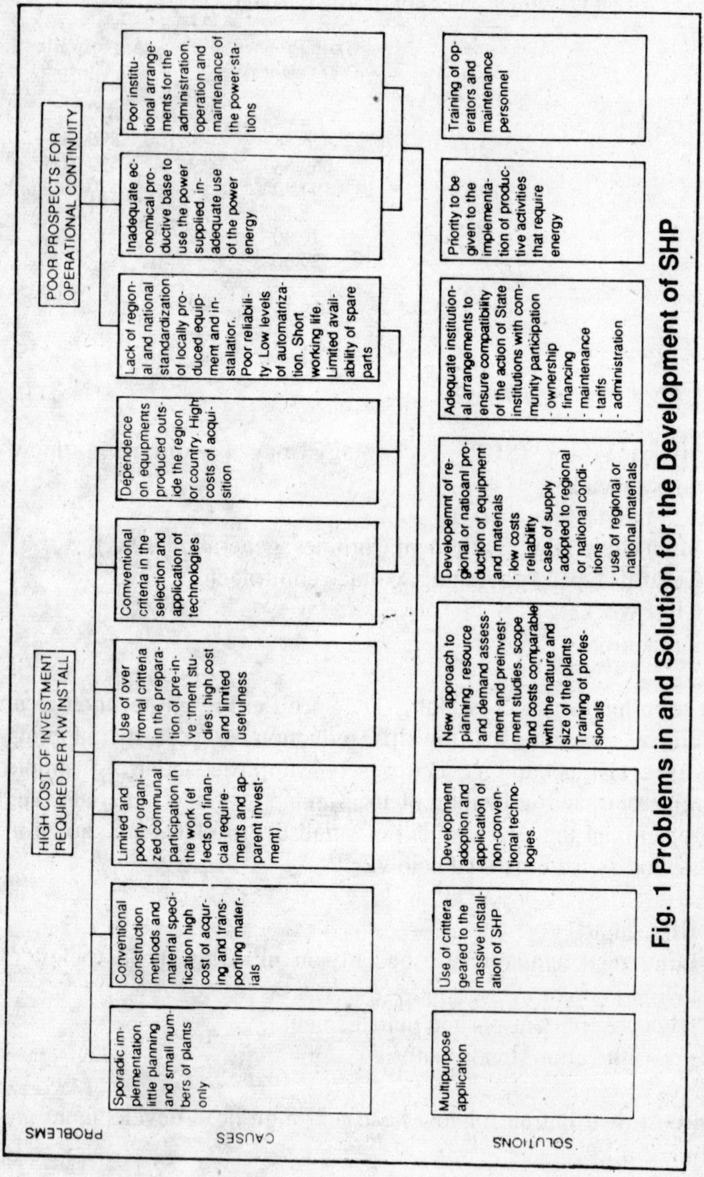

Fig. 17.1: Development of Small Hydro-plants.

Hydraulic Turbine

Types of turbines suitable for small hydro development and some manufacturers are shown in Table 17.2. Recent developments in turbine technology are standardized units. Ultra low head (below 3 metres turbine and adoption of pumps for use as turbines in reverse mode to reduce costs.

Generators

Conventionally hydro-electric alternating current synchronous generators are used. The generator (rotor) should withstand turbine runaway speed, i.e. the speed attained on sudden load throw-off. Brushless excitation system generators used in mini hydel system can reduce maintenance costs and time. Induction generators can also be used. Induction generators need excitation source and as such can operate only when connected to an existing electrical system from which it can obtain its excitation current.

Controls and Protection

Flow control by speed governing and other controls in mini-hydro plants form a significant component of equipment costs. Further controls need attended operation. An electronic controller to keep the load constant and to reduce costs and simplify operation has been developed. Simplicity, adequacy within economical limits is the major requirement in evolving the control system.

Civil Works

Civil works are highly site specific. In a high head plant, typical elements of a small mini plant are diversion and intake, desilting tank, channel, forebay, penstocks and power house. A typical[4] high head scheme is shown in Fig. 17.3.

For economic exploitation of small hydro power civil engineering components of power station such as diversion dam, head race channel, sand trap, forebay and power house building should be simple and cheap for construction and maintenance. Utilizing existing facilities saves the civil works costs considerably.

The simplest form of a small power station is run-of-the-river plant with no storage dams. A run-of-river-plant has an indirect benefit in case of competing uses such as irrigation and flood control, for the water that is used to generate hydro-power. An engineering estimate of the flow should be made for micro hydro-power generation in the absence of stream flow records on the basis of local enquiry and observed data to determine the adequacy of water and to fix the size of generating units.

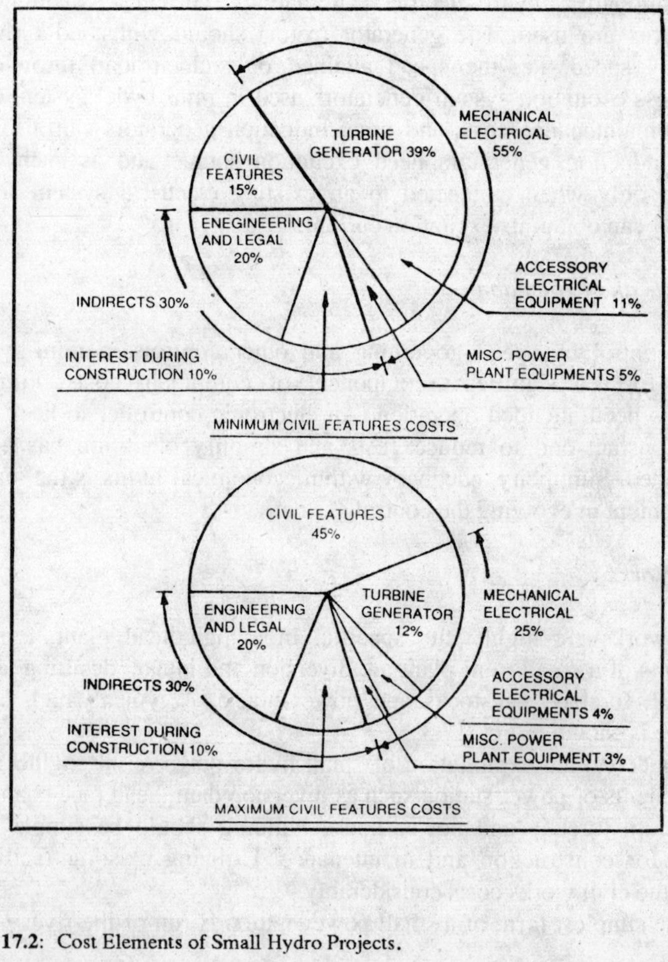

Fig. 17.2: Cost Elements of Small Hydro Projects.

TABLE 17.2: **Type of Turbine for Small Hydro development**

Type of Turbine	Names of some of the manufacturers	Suitable for		Remarks
		Head in metres range	Flow in cubic metres range	
1	2	3	4	5
1. Cross flow Turbine (Banki-Mitchell or osberger Turbine)	1. Osberger Turbines Fabric Federal Republic of Germany 2. Balaju Untrashala, Nepal	1 - 200	0.03 - 9	
2. Axial flow				
(a) Stra-flow	1. Eischer Wyss, Switzerland 2. Osta Chalskara, USSR	2 - 50	3 - 20	
(b) Tube	1. All is Chalmers, United States 2. J.M. Voith, Federal Republic of Germany 3. S. Rumsand Venksted, Norway 4. Boving, United Kingdom 5. Jyoti Ltd., Baroda (India) 6. Bharat Heavy Electricals, Bhopal, India 7. Flovel, Faridabad (India) 8. Voiest Alpine, Austria	2 - 15	3.5 - 30	

Contd.

TABLE 17.2 *Contd.*

1	2	3	4	5
c. Bulb	1. Neypric, France			
	2. Eischer Wyas, Switzerland	1.25-25	3 - 70	
	3. J.M. Voith, Federal Republic of Germany			
	4. Fuji Electric, Japan.			
	5. Tampella, Finland			
3. Francis	1. Voist Alpine, Austria	8 - 200	0.3 - 6	
4. Kaplan (Vertical)	2. Tampella, Finland	1.1 - 70	3 - 40	
5. Pelton	3. Dress & Co., Federal Republic of Germany	45 - 1000	0.06 - 3	
	4. Flovel, India			
	5. S.Rumsand Venksted, Norway			
	6. Jamco Leffel, United States			
	7. J.M. Voith, Federal Republic of Germany			
6. Turgo Impulse	1. Jyoti Ltd., India			
	2. Gilbert, Gilkes & Gordon, United Kingdom	45 - 300	1 - 8	
7. Schneider Hydro Engine	1. Lotte Schneider Engines, South Korea	0.5 - 15	1 - 8	
	2. Schneider Engine Co., Justin, Texas	5 - 30	0.05 - 3	
8. Pump as Turbine				

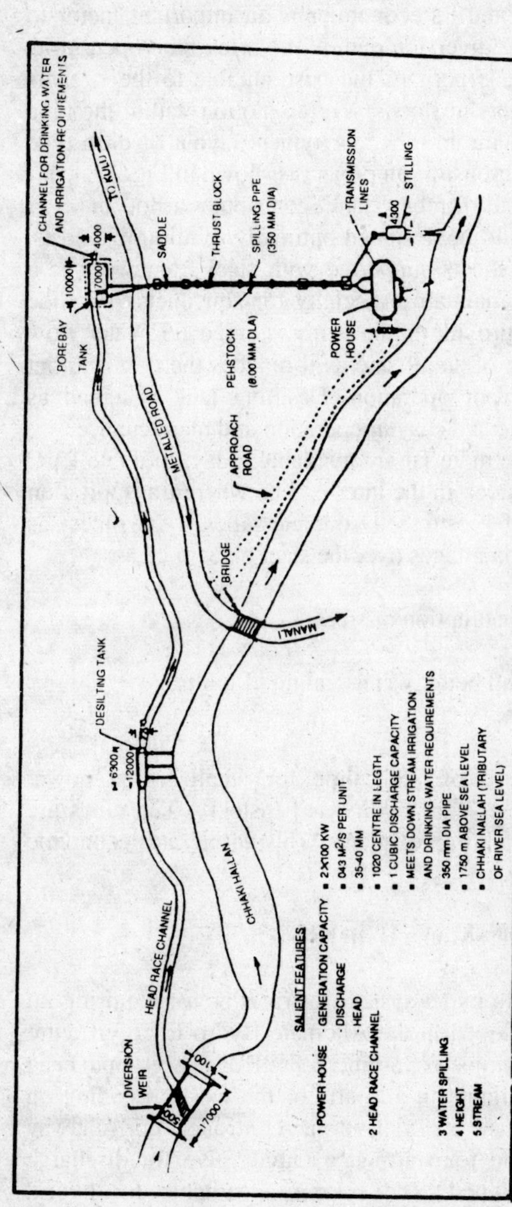

Fig. 17.3: General Layout of Manali Micro Hydro Hybrid Energy System.

Diversion dam design and its economics is an important factor to be considered. A drop type diversion dam with vortex tube open in its top, placed in stream bed, is perhaps the best suitable to the streams carrying large-size sediments in floods. Vortex motion within the tube prevents lodging of sediments in tube.[4]. A typical layout of drop type vortex tube proposed in demonstration project is shown in Fig. 17.4.

Head race channel and forebay tank contribute a lot in civil engineering costs and should be designed optimally to minimize costs. Channels made by metal sheets supported with steel framework is a cheap option and easy to maintain especially for aqueducts and slide zones. Forebay tank can provide a short-time storage for water flow fluctuations, but should be of small size as it reduces the cost without much trouble in power plant operation. Desilting tank designed as vortex type is cheap, simple and easy in operation and maintenance.

Various options in pipe material are available today. Suitable PVC pipe could be used as penstock in the interior area where transportation costs a lot. Application of prestressed concrete pipes (PSC pipes) as penstock has got several advantages over the steel pipes such as:

- Low cost and low consumption of steel,
- More durable,
- Easy in installation and better water sealing of joints,
- Low maintenance cost.

In China extensive use of PSC pipe for small hydro power installations demonstrates its adaptability as penstock.[1] The pipes are made at a central site for a group of plants. This accelerates economic and industrial development.

New Decentralized perdeep Micro Hydel System

A new economically viable hydel system for rural areas, requiring no attendance, has been developed in the Alternate Hydro Energy Centre (AHEC). Conventional pumps as turbine for all discharge and head conditions are readily available in all parts of the world. Loading on the generator is kept constant by an electronic controller. For runaway condition under failure and for starting, a radial valve for discharge throttling has been developed. The system is suitable for hybrid development. For details refer a recent paper.[5]

Induction generator coupled micro-hydel working in integration with biomass gasifier operated synchronous generators for governing forming local power distribution network with an emergency link with the main power system is conducive to provide an economically viable system, meet energy needs and provide substantial energy input for main grid.

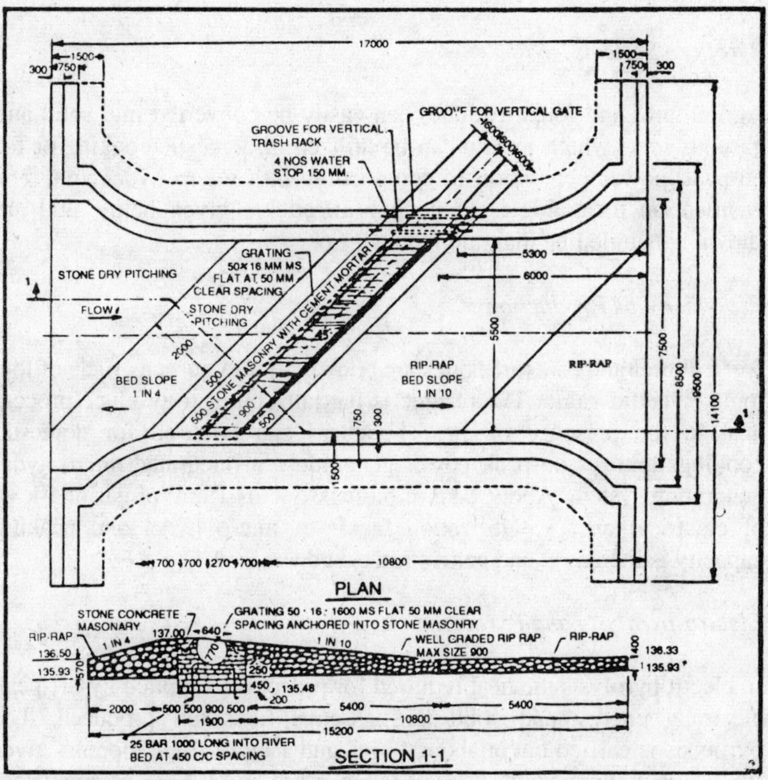

Fig. 17.4: Drop Type Vortex Tube Diversion Dam at Manali.

Hybrid System Development and Utilization

A major cause of low benefit cost ratio of isolated small plants is very poor utilization of power generated from flowing water. This can be overcome by using surplus off-peak high grade electrical energy to upgrade other forms of energy in hybrid systems. A typical block diagram of such system is given in Fig. 17.5. Agro-waste and forest residues are converted into solid and gaseous fuels, liquid fuel alcohol, proteinous animal feed, liquid nitrogen fertilizer, etc. decentralized integrated hybrid systems consisting of micro-hydel, biomass gasifier connected generating units and other systems are economically viable, stimulate industrial and economic development activity. Some of the hybrid systems are briefly discussed.

Pyrolysis Systems

Agricultural and forest residues can easily be converted into solid and gaseous fuels which in turn can be utilized in domestic cooking or for on-peak power generation in conjunction with micro hydel unit. The gasification technologies to be hybridized are given below and are shown by single line diagram in Fig. 17.6.

Producer Gas Production

Agricultural and forest residues are converted into gaseous fuels of low joule calorific value. The residue is partially burnt to produce process heat in the presence of air. The gases can be used for domestic cookings and for on-peak power generation in integrated micro hydel generation system. About 1 kg of biomass yields 1 nm^3 of gas or 1 kwh of electrical energy. Utilization factor of micro hydel and peaking capacity is increased and generation cost decreased.

Electro-pyrolysis and Pyrolysis

In electropyrolysis the heat required for process is supplied by off-peak electrical energy and solid and gaseous fuels are produced. The pyrolysis is carried out at about 400°C and 1 kg of dried biomass gives 0.5 kg of medium joule heating value gases and 0.5 kg of smokeless solid fuel. The gas can be utilized for supplementing micro hydel generation capacity and distributed through a pipeline network as domestic cooking fuel. The process requires heating the biomass off peak surplus micro hydel electrical energy. Following reaction takes place in the pyrolysis:

$$\text{Biomass Heat} \quad CO + H_2 + CO_2 + \text{Higher hydrocarbons} + \text{Char} + \text{Tar.}$$

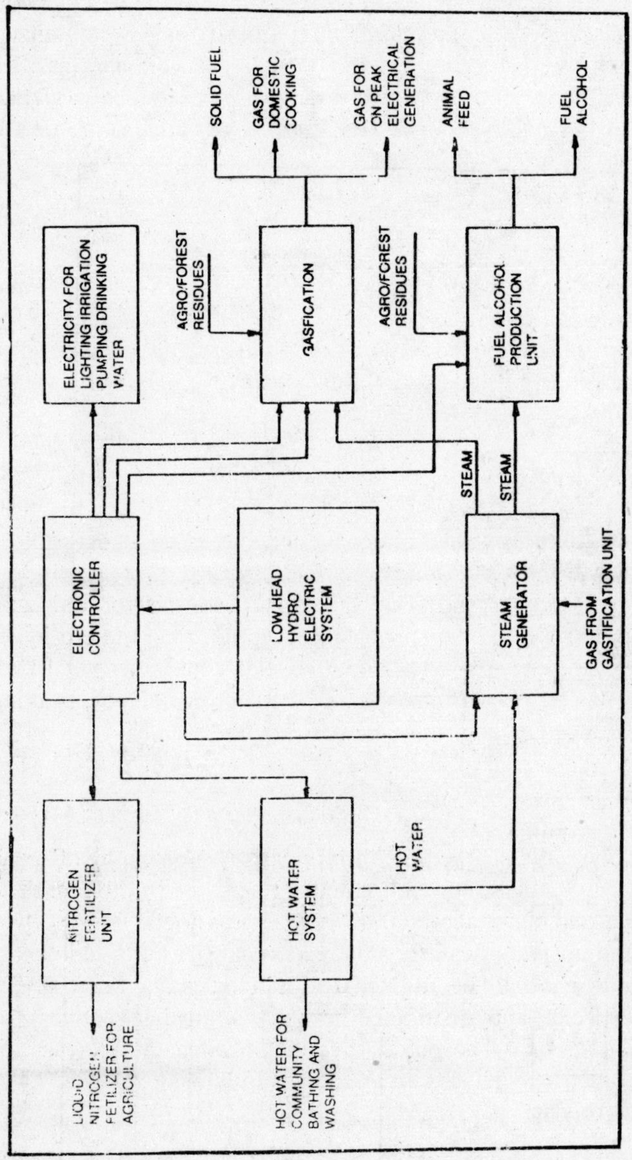

Fig. 17.5: Block Diagram of Hybrid System.

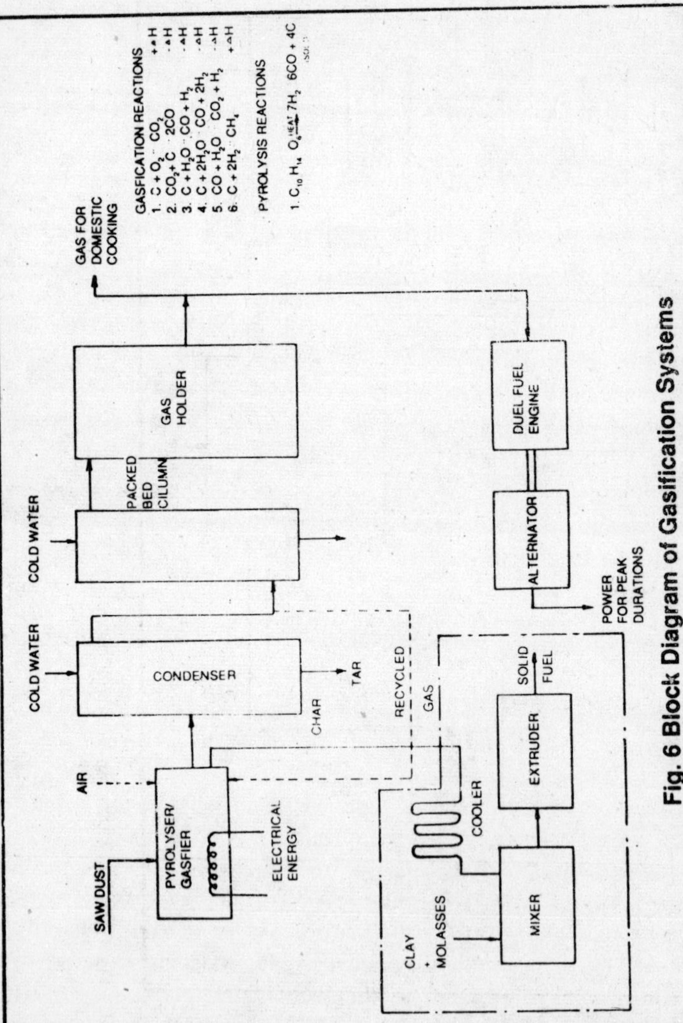

Fig. 6 Block Diagram of Gasification Systems

Fig. 17.6: Block Diagram of Gasification Systems.

Tar is separated by condensing and the gases are further cleaned by scrubbing with water and sent for power generation or as pipeline gas. Hot char is cooled by water and then mixed with clay and molasses in the mixer. The mix is fed to the extruder for producing briquettes of desired size.

In the pyrolysis the process heat is supplied by burning 20 per cent of the pyrogases generated in the system. Thus the process is started with electrical energy and subsequently the gases can be used for providing energy to the system. The remaining gases after purification are used for power generation and the char for producing solid fuel.

Production of Animal Feed and Fuel Alcohol

Biomass can be upgraded with electrical energy for production of high protein value animal feed and fuel alcohol for blending with gasoline/diesel. The process flow diagram is shown in Fig. 17.7. It consists of first hydrolysing the lignocellulosic material and breaking of lignin and hemicellulosic structure for proper utilization of mass to the enzymatic decomposition. The cellulose is decomposed into glucose which can be used as animal feed or for further fermentation into alcohol. The glucose is fed to the fermentor where it decomposes into ethanol and carbon dioxide gas. Dilute alcohol is distilled to produce 190 proof alcohol to be used for blending with gasoline. One kg of biomass yields about 1/2 kg of animal feed or 1/4 kg 190 proof fuel alcohol.

Liquid Nitrogen Fertilizer Production

The local nitrogen fertilizer requirements can easily be met with the production of liquid nitrogen fertilizer (calcium carbonate) by the electric arc process in integration with micro hydel system. The process requires only calcium carbonate/rock phosphate, air, water and electrical energy. The flow diagram of the unit is given in Fig. 17.8.

The basic equipment required consists of ballast transformer, air and water pump, reaction cell, made of 0.08 m diameter mild steel pipe with a concentric electrode at the centre. Arc is formed by supplying 3,000 volt D.C. between the concentric electrode and reaction tube body. Air is tangentially fed to the tube which gets converted into nitrogen oxide and is absorbed in water and calcium carbonate/rock phosphate. The conversion efficiency depends upon the length of the arc formed and the amount of the air passing through high temperature plasma zone. The reaction tube is thus designed to produce vortex type arc and cooled externally by circulation of cold water. A 3 KW unit is capable of supplying the nitrogen fertilizer needs of 5 to 20 hectare land depending upon the amount of fertilizer applied per hectare.

Economic Evaluation

The economic evaluation of the micro-hydel generation system in integration with biomass energy for supplementary generation is shown in Table 17.3. The production cost per unit of electricity in conventional micro hydel system comes to Re.1 while when integrated in hybrid system the cost is reduced to only Re.0.34-0.41 which is very competitive with prevailing prices. Also the hybrid option produces briquetted fuel for domestic cooking.

TABLE 17.3: **Economics of using Combined and Integrated System**

(Micro Hydel 100 kw)

Sl. No.	Installation Type	Installed (peak capacity) kw	Annual Plant Factor (Micro hydel) (in percentage)	Average Cost per per unit (in Rs.)
A.	*Decentralized (Isolated)*			
1.	Micro hydel	100	40	1.0
2.	Integrated New Micro Hydel (100 kw) Gasifier Unit (100 kw) (peaking)	200	55	0.49
3.	New Integrated Micro hydel (100 kw) Electrical Gasifier/ Briquetting (100 kw)	200	60	0.41
B.	*Grid Connected*			
4.	New Micro Hydel (100 kw) Gasifier/Briquetting (100 kw)	200	80	0.34
5.	Grid connected New Micro Hydel	100	80	0.40

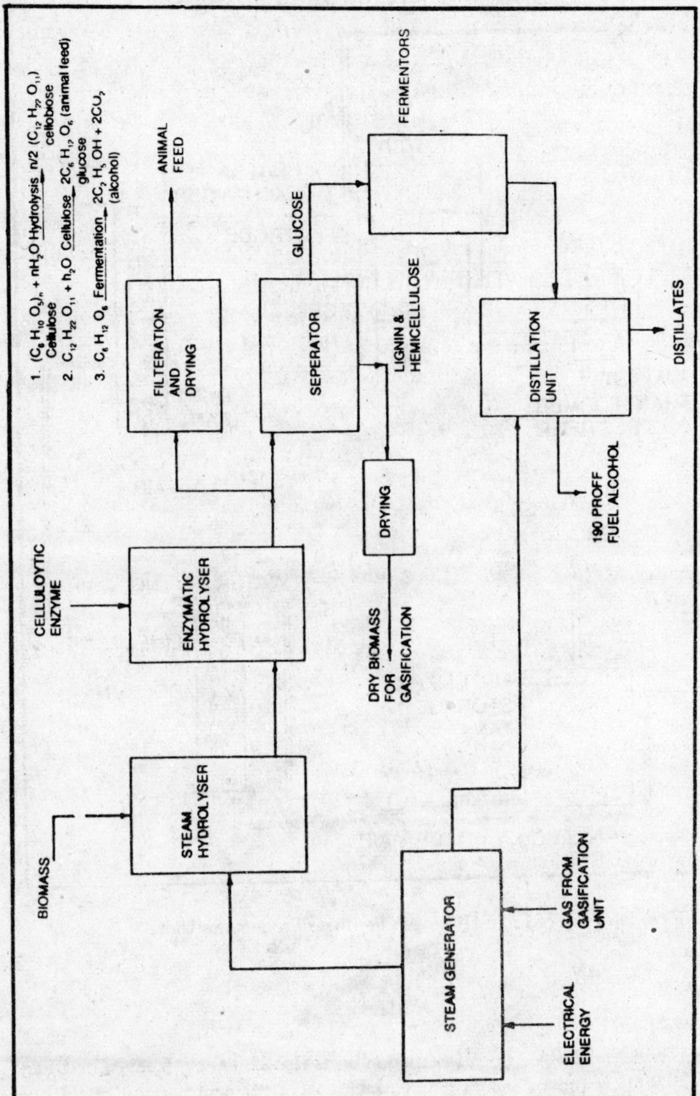

Fig. 17.7: Flow Diagram of Animal Feed and Fuel Alchol Production.

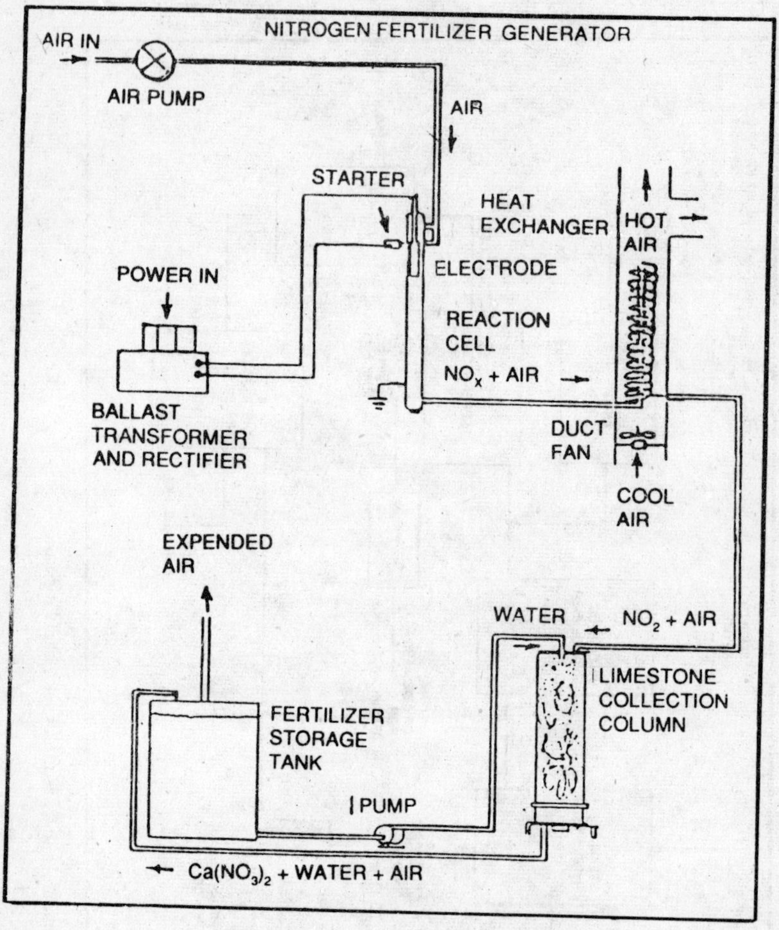

Fig. 17.8: Flow Diagram of Liquid Nitrogen Fertilizer Productioin Unit.

NOTES

1. Micro-hydel cost based upon installation costs Rs. 25,000/kw (conventional) and Rs. 20,000/kw (new); annual maintenance cost 2 per cent, interest rate 10 per cent, depreciation 2 per cent and 6,000 annual working hours.
2. Gasifier cost based upon installation cost of Rs. 8,000/kw, annual maintenance cost 3 per cent, interest 10 per cent, depreciation (average) 4 per cent and Re.0.5/kg biomass cost.

3. Gasifier-cum-briquetting fuel plant cost based upon installation cost of Rs.8,000/kw, annual maintenance cost 3 per cent, interest 10 per cent, depreciation 4 per cent, Re 0.5/kg biomass cost, Re 1.0/kg selling price of smokeless briquetted fuel and labour cost Rs. 10/hr.
4. Average load factor 40 per cent.
5. Briquetted fuel produced as by-product at the rate 15 kg/hr.

REFERENCES

UN ESCAP ECDC-TCDC *Directory on Renewable Sources of Energy*, Vol. IV "Small Hydro Power Development," 1982.

India - 1982, Publications Division, Ministry of Information and Broadcasting, New Delhi.

Thapar, O.D., "Hydro-Electric as a Future New Energy Source", 3rd World Congress on Water Resources at Maxico, 1979.

Thapar, O.D. and Arun Kumar "Integrated Small Hydro Energy Development", ASCE Symposium, 1984.

Thapar, O. D., Joel B. DuBow, "New Micro Hydel System," 4th Miami International Conference on Alternate Sources of Energy, Miami.

Thapar, O.D., "Decentralised Mini/Micro Hydro Energy," International Training Programme Committee on Science and Technology in Developing Countries, Madras, 1984.

Parikh, P. P., J. A. Pandya and P. Arikal, "Biomass Gasification — An Appraisal and Viability Assessment", *Urja* No.4, 1985.

SESSION V

ENERGY MANAGEMENT AND CONSERVATION

Chairman: Prof. K. L. Chopra,
Indian Institute of Technology, New Delhi

EIGHTEEN

SOLAR THERMAL SYSTEMS FOR POWER GENERATION

PROF. M. RAMAKRISHNA RAO

Solar energy is a renewable source of energy which is in abundance in nature. The solar energy falling on India is about 10^{16} Kwh per year. If we could tap 0.1 per cent of this power by the next century, our energy requirement could be met adequately. So, R and D efforts in harnessing solar energy is of paramount importance to India as well as to other countries when the fossil fuels are depleted every year. Already these efforts that have been put in, during the last two decades have resulted in the establishment of solar thermal and photovoltaic systems on a commercial scale in USA, Israel, Europe and Japan.

Solar thermal energy systems, which use concentrating solar collectors to convert the Sun's radiant energy into heat energy, can contribute significantly to nation's future energy needs. For low temperature applications, the flat plate collector systems are used for water heating and air drying systems. But the major emphasis in solar thermal technology has been on three concentrator system concepts viz. the central receiver, the point focussing parabolic dish and the line focussing parabolic trough. Applications of these systems include the production of electricity by central generating plants and smaller distributed plants; production of process heat for industrial and agricultural uses; and congeneration where the energy is used to produce both electricity and process heat.

The central receiver system solar I, the 10 MW USA Government funded plant is in operation in California. The private agencies are proposing to construct 100 MWe and 30 MWe plants at a cost of

\$3,200 per KW by developing the low cost heliostats. The other concept of distributed receiver technology is gaining momentum as the small plants of low capacity could be installed for individual industries and small community requirements. In this category ten parabolic dish systems and eight parabolic trough systems are installed in USA at user sites by 1983 for power generation and industrial process steam. The notable installations are 200 parabolic dish system by La Jet Energy Co at California to produce 4.4 MWe at \$3,790 per KW and Luz Company installation of 13.8 MWe parabolic trough system at \$4,000-4,500 per KW.

With this background, let us examine the Indian scenario in solar thermal energy systems. Even though the importance of this new energy resource was realised as early as in 1975, the first proposals fructified in 1981 with Indo-USA collaborative programmes in solar energy. The two important projects are power generation by parabolic dish concentrating system as a part of the rural integrated energy programme at Salojipally, Medak district, Andhra Pradesh, and industrial process steam by parabolic trough concentration system at Government Silk Factory, Mysore, Karnataka. Both projects were sponsored by Department of Non-conventional Energy Sources (DNES), Government of India and USAID of US Government. The Principal Investigators are from Bharat Heavy Electricals Ltd., Hyderabad, and Central Electronics Ltd., Delhi, on Indian side and from Jet Propulsion Laboratory on US side for dish system and Indian Institute of Science, Bangalore, and UNIV of Houston for line system.

Under the parabolic dish project, it is proposed to fulfil the energy needs for irrigating the dry land in the close proximity to the Salojipally village. The electrical power capacity is designed to provide about 25 KWe (peak) power which will drive at least 8 numbers of 3 KW motor pump sets, irrigating 32 acres. It consists of 6 heliodishes of 9 m dia each covering about 50 m^2 collector field. The fixation and alignment of 225 mirrors in each dish and receivers has been completed. The heat transfer fluid is water and the concentrators produce super saturated steam at 550°C at 80 atmospheres pressure. The flow rate is 180 kg/hr. The reciprocating steam engine is utilised for conversion power system. The collector field is driven by microprocessor based tracking two directional system. The project is nearing completion and in a couple of months the system will be commissioned. The other project is to design and develop a medium temperature solar line focus concentrator system for steam generation at 150-180°C initially for industrial process application. The principal investigators are from Indian Institute of Science, Bangalore and University of Houston, U.S.A.

The collector field consists of 30 parabolic trough concentrators each of 6 m^2 area arranged in five rows. Six modules are connected in series in each row and are driven by a gear reduction motor to track the Sun. The concentrators are designed with rib structure which are light weight and cost effective. The reflecting surfaces are flexible, thin, back silvered mirrors (0.4 mm thick), sandwiched by a thin (0.4 mm) steel sheet with epoxy. The receiver tube is 30 mm steel tube coated by selective black chrome developed at Indian Institute of Science, Bangalore. A micro-processor based flux line tracker system is used to control the system. A flash boiler system is used to convert the pressurised water in the concentrator receiver tube raised to a temperature of 180°C at 10 kg/cm^2 pressure to 4.8 kg/cm^2 at 150°C. The steam thus generated is fed to the silk printing tables in the printing section of Government Silk Factory at Mysore. At present, the steam is generated by a conventional coal fired boiler. This solar system will be interfaced with the conventional boiler steam supplying line.

The collector field is already installed and the plant will be commissioned in a couple of months. This produces 708 KW thermal power and steam of 800 kg/day at 150°C. A similar collector field could be utilised to produce steam at 250°C to drive a turbine to generate power.

In addition to these two national projects, DNES has proposed to instal a demonstration plant of 50 KW solar power plant for Achheja village near Delhi. The contract is being awarded to M/s. MAN of West Germany and Coppus of U.S.A. The basic functions of this plant would be to energise irrigation pumps, provide power for lighting, energise threshers, lathes, provide energy to village level medium industry and operate cold storage/milk chiller unit etc. This plant consists of 620 m^2 collector area of parabolic trough concentrator type coupled with Coppus steam turbine capable of working between 25-75 KW produces 25 KWe power. The efficiency of the concentrators is about 70 per cent and that the turbine is about 20 per cent. In case the outlet steam pressure is maintained at 10 PSIG in the turbine, the steam consumption per hour shall be 1,320 kg/hr for power generation 18 KWe. However, the waste heat of turbine outlet steam could be utilised to operate a refrigeration system of 6 tons capacity for operating a chilled/cold storage plant. In the phase-II, the collector field would be added to generate more power up to 50 KWe. The total cost of the plant is about a million dollars. The power generation by solar pond technology is under progress at Indian Institute of Science, Bangalore, and at Baroda through Gujarat energy development agency.

The survey for central receiver system for 5 MW power is being done by DNES and will be taken up in VII plan. The rich experience of other countries, the fast advancement in the technological developments and collaborative programmes have accelerated the progress in the implementation of renewable energy programmes in India. It is hoped that the coming years would see more and more achievements in high technology related solar energy programmes.

NINETEEN

ISSUES IN ENERGY PLANNING

ASHOK AIMA

INTRODUCTION

At present issues and problems pertaining to the energy sector seem to have come under the focal attention of planners in the country. A sudden consciousness has dawned on the planners to have a suitable and comprehensive energy policy. In fact, the absence of an adequate National Energy policy earlier has resulted in massive depletion of our natural resources. The burden of catering to the growing energy requirements of our huge masses coupled with the staggering demand generated by the developmental pursuits has been so heavy that our national forest wealth has depleted at more than 1 per cent over the last couple of decades. Recent sattelite observations reveal that at present the land covered by the forests stands at a much lower level than the popular estimation of it being around 23 per cent. The growing demand for firewood and timber in the predominantly rural agrarian economy like ours, is also putting pressure on our forests which are the main source for catering to energy requirements. Though the recent oil finds have been quite substantial in India yet it has not affected the consumption pattern of the rural sector to a significant degree – it remains the same as it has been for the last so many decades. Now the planners are not only worried about as how to meet the growing demand for energy but they are faced with the dilemma of escalating ecological imbalances in the economy as well, which is compounded by the ever increasing demand for timber. Thus these problems have necessitated an elaborate and comprehensive National Policy with the twin objective of meeting the growing demand for energy products and also of warding off any future adverse impact on the eco-system which generally is the consequence of indiscriminate deforestation.

OBJECTIVES

The present study pertains to the inhabitants of the Dachigam-Telbal catchment area. The catchment area has been brought under the "Eco-Development Project" by the Department of Energy, Government of India, for restoration of its ecology. The 420 households were surveyed through stratified random sampling technique based on the personnel interview method by using a prepared questionnaire. An attempt has been made to study the energy consumption pattern of the catchment areas so as to study the following objectives:

1. To study the geographical setting of the catchment so as to gain an insight into the topography of the region;
2. To study the income and employment pattern of the catchment so as to establish its relationship or otherwise with the level of energy consumption;
3. To study the energy pattern of this region so as to assess its impact on the local forests; and
4. On the basis of the results drawn from above information, policy measures would be identified for the restoration of balance in the region's ecology.

Location of the Catchment

The Telbal-Dachigam area represents a typical rural entity where the process of urbanization has been quite significant, being on the outskirts of the main city centre of Srinagar. This region comprises 25 villages with a total population of 38,457 persons (21,530 males and 16,927 females) living in 4,215 households (Census, 1981). Out of these 25 villages 16 villages with 2,397 households are compact, while four villages with 478 households are semi-compact. The remaining five villages with 1,340 households are scattered (See Annexure-I). Of the total village settlements 15 villages are located in the plains of the region, three villages are hilly and seven villages are undulating. The total population of the sample households is 2,514 persons (1,321 males and 1,193 females) giving a sex ratio of 903 females per thousand males. In other words 52.55 per cent of the sampled population is male and the remaining 47.45 per cent of female. Out of a total sampled population 42.36 per cent are minors and 57.64 per cent constitute adult population age group. The average size of the households is 7.75 persons with a population density of 507.40 persons per sq. kilometre in the catchment.

Employment Structure of the Catchment

The employment pattern of the sampled population reveals that 25.17 per cent of the total population consists of earners and 74.83 per cent is economically dependent. Employment structure (See Annexure-II) of this catchment shows that agriculture and its allied sectors employ 37.65 per cent of the working population, followed by business sector employing 20.06 per cent of the work force, Government sector employees 11.11 per cent of the working force, while daily wage earners constitute 16.17 per cent of the active population. Needlework, poultry, fishing and others constitute 5.75 per cent, 4.94 per cent, 3.09 per cent and 1.23 per cent of the overall working population respectively. It is amply evident that nearly 78 per cent of the working population draw their subsistence from and within this catchment, supported by the Government Sector. It can thus be observed that the brunt of employment generation over the period is still being borne by this rural entity itself. However, one noticeable feature which emerges from studying the existing employment behaviour in the catchment is that the agriculture sector's role of being the mainstay of the people is gradually diminishing. Household industry with horticulture and animal husbandry practices are growing at a significant pace as against the growth in the agricultural sector.

Income Pattern of the Catchment

The per capita annual income has been found out to be Rs.1,648.08 in the catchment amongst the sampled households. The respective share of various sectors (See Annexure-III) to overall income generation shows that agriculture contribute just 2.65 per cent to it, household industries contribute 17.73 per cent, while Government Sector contributes 18.19 per cent to the income generation of this catchment area. Daily wage earners and business sector contribute 5.2 per cent and 11.16 per cent to the overall income generation respectively. Livestock trade with horticulture and household sector emerge as the main income contribution in the catchment with 35 per cent, 10.07 per cent and 17.73 per cent respectively. Thus it becomes clear that the role of the agricultural sector as the bread earner is insignificant in the overall economic scenario of the catchment.

Energy Consumption in the Catchment

Broadly speaking the energy consumption (See Annexure-IV) pattern of this catchment area can be categorized into following items:

1. Firewood
2. Soft-coal
3. Agricultural waste
4. Electricity and
5. Kerosene oil and Petrol

Firewood

Of the total firewood requirements in the catchments only 39.72 per cent is procured from the market as 61 per cent of the firewood is obtained through domestic and indigenous sources. The average household monthly consumption in the catchment area is Rs. 99.92 of which domestically procured component is Rs. 60.95 and market purchase component Rs. 38.97. The source for firewood is the local forests. Nearly 419.60 qtls of firewood is consumed monthly among the sampled households in the catchment area and the entire supply of this firewood comes from the local forests. On an average if one tree can yield 10 quintals of firewood it can be safely said that at least 42 trees are felled every month for meeting the energy requirements. It, therefore, means that 420 trees are chopped off every month to meet the energy requirements of the entire catchment. Firewood consumption accounts for 39.70 per cent of the overall energy consumption in the catchment. This high percentage of firewood consumption reveals that this catchment has predominantly primitive and traditional energy consumption pattern. Firewood is mainly used for cooking purposes, i.e. morning tea and breakfast, mid-day meals, evening tea and dinner. To some extent firewood is also used to heat water for washing and bathing purposes during the prolonged winter months.

Soft Coal

The next major item of energy consumption is soft coal. This coal is used in *Kangris* (a traditional type of body heating equipment) which is used for nearly seven months in a year, i.e. from late September to April. In the absence of adequate heating arrangements in the houses, people make frequent use of *Kangri*. Usually one quintal of wood yields two quintals of soft coal. In overall energy consumption basket of the people in this region 27.03 per cent is accounted for by the soft coal consumption. However, soft coal being the residual of firewood, nearly 81.74 per cent of this coal consumption is met by the firewood already used for cooking purposes and remaining 18.26 per cent of the soft coal is purchased from the local market. Again the source of the purchased soft coal is the coal forests, and as such nearly 80 to 100 trees are chopped off every season to meet the local coal requirements of the sampled households. Practically 1,000 trees are burnt off per annum to meet the energy requirements of the entire catchment. Thus the problem of deforestation is further compounded on account of soft coal requirements as well.

Agricultural Waste

This type of energy is used in this catchment area to cater to cooking purposes of the households in the catchment. The source of this waste is the farm and horticulture land which supplies the entire quantity of this energy product and is entirely met through domestic sources. The residue of this burnt out agricultural waste is used as local manure on the farms as well. Agricultural waste accounts for 21.35 per cent of the total energy requirements of this region. The role of the agricultural sector as source of energy is not as prominent as one tends to envisage in a typical rural oriented society. It is because its contribution to overall economic activity is quite modest. The horticultural sector of late has lent some degree of credibility to agriculture as it is becoming one of the popular sources of economic activity. One tends to observe that the traditional farm land is now being increasingly brought under fruit orchards. Thus, as a source of energy, agriculture on its own has a limited role as the source of supply of energy requirements. However, one redeeming feature is that deforestation is not the consequence of growing horticultural practices, which, on the contrary, have helped to restore some degree of balance in the catchment ecology.

Cowdung

Another source of energy consumption in the catchment is cowdung which is used for cooking and heating purposes by the people. The main source of this energy is the livestock sector. A noticeable feature of the study of the energy consumption pattern is that though livestock accounts for 35 per cent of the total income in the catchment, its contribution as a source of energy is just 6.73 per cent in the overall energy supply. The entire quantity of cowdung is domestically procured. The residue of this burnt-off cowdung is also used as local manure in the agricultural operations. Thus cowdung serves the twin purpose of an average village man, i.e. he uses it as a source of energy and also as a fertilizer.

Electricity

Electricity accounts for just 2.78 per cent of the total energy requirements of the sampled households in the catchment. This source of energy is used mainly for lighting purposes. Electricity has a restricted role in the energy consumption pattern as it is neither used for cooking nor for heating purposes. This explains its insignificant role in the overall energy requirements of the region. Though the State as a whole has a vast proven hydro-electric potential yet for most of the winter months it is a net importer of electricity from other States. It is hoped that with the completion of on-going hydro-electric power stations, the State will be able to provide substantial electricity to the consumers. This may as well lead to a shift in the energy area. Thus, in future electricity may replace the traditional source of energy which in turn may help to check the deforestation problem.

Kerosene oil and Petrol

This source of energy is the least consumed article in the energy consumption basket of the people in this region as it accounts for just 2.39 per cent of the overall energy consumption. The main reason for the low consumption of kerosene and petrol is because an average household in the region is dependent on the traditional source of energy, i.e. firewood and cowdung under the *Chulla* system. Kerosene oil is used in the stoves and lamps for cooking and lighting purposes. Few of the households own cars and scooters which require consumption of diesel and petrol.

Conclusions and Suggestsions

The area under study represents a typical rural entity where people are predominantly dependent on the local catchment forests for their overall energy requirements. With a rapid increase in the population in this region, fuel requirements have also increased many fold. The breaking off of the joint family system has swelled up demand for energy. The prevalence of traditional methods of energy consumption is responsible for the low fuel efficiency, which necessitates a much higher level of energy consumption to produce the requisite degree of heat. As such it can be inferred that the pressure on the local forests for providing means of energy to the growing population of the area is tremendous. Consumption of petroleum-based energy seems to be abysmal which demonstrates that this region has still not taken to "Petro-Energy". The average monthly energy consumption of a household in this catchment works out to Rs. 251.50 at present (Rs. 151.50 is the domestic component while Rs. 100 is the market purchased). In fact, except for smaller quantities of soft coal imported in to the region, rest of the firewood and soft coal requirements are catered to by the local forest. Unless some drastic changes are make in the existing pattern of energy consumption, the growing population of the catchment will have a disastrous effect on the local forests. This would not only result in rapid denudation of forests in the catchment but will also cause soil erosion on a massive scale creating silt problems in the adjoining Dal Lake, as the tributaries of the catchment run into it.

Given the existing energy endowments in the region some policy measures can be suggested based on the above discussions:

(i) There are good chances of installing bio-gas plants in the catchment, because the region is quite rich in the prerequisites for bio-gas. Not only the farm waste can be used for generation of bio-gas but rich cattle population of the region can also augment the source of supply. Human waste can also be used for this purpose supplemented by growing horticultural waste. However, at present the existing type of bio-gas technology cannot be straightaway applied in this region. An appropriate technology will have to be evolved for the introduction of bio-gas plants in this region,

(ii) Apart from developing region specific bio-gas technology, attempts need to be made to make the *Chullah* system more efficient. Therefore, the Chulla should be redesigned and improved upon to increase its efficiency as it would lead to using lesser amount of time for making day-to-day meals,

(iii) Kerosene stoves could also be used during the summer months in this region which would ease out the pressure on the demand for firewood. For this the Government could subsidise the cost of stoves and of kerosene so as to make it an attractive proposition to the people in the catchment, especially to the lower income group. Keeping in view the cost of afforestation the official agencies would have to bear and the time period involved in restoring the balance in the local ecology a subsidy on kerosene would not only prove a short-term palliative but also would help in changing the energy consumption pattern of the people of the region,

(iv) A commercial forest zone could be created in the catchment, wherein fast growing variety of trees could be planted to meet the firewood requirements of the region. Subsequently a rotation system in the plantation system of those fast growing trees would help to check the depletion of the forest wealth, and

(v) Cooking utensils which get quickly heated-up could also be designed and popularized. This would result in higher fuel efficiency as well as saves considerable amount of labour involved in day-to-day cookings. This again would reduce the quantum of firewood used for cooking and heating purposes. Extension agencies at the block level have to make concerted efforts to popularise these steps.

The need for a well-defined integrated and coordinated energy policy has become imperative to avert the disaster which has engulfed our national forests resulting into a massive disequilibrium in our eco-system. Thus policies so designed would also have to be compatible with the diverse and varying requirements of the difficult regions of the economy.

ANNEXURE I

Locational Setting of Telbal and Dachigam Catchment

		Villages	Households
a.	Type of Villages		
	Compact	16	2397
	Semi-compact	4	478
	Scattered	5	1340
		25	4215
b.	Location of Villages		
	Plain	15	2956
	Hilly	3	732
	Undulating	7	527
		25	4215
c.	Accessibility		
	Bus service available	22	3492
	Bus service not available	3	723
		25	4215

ANNEXURE II

Employment Structure of Telbal Dachigam Catchment

	Occupation	Percentage of Labour Force
1.	Farming	37.65
2.	Business	20.06
3.	Daily casual labour	16.17
4.	Government Sector	11.11
5.	Needle work	5.75
6.	Poultry	4.94
7.	Fishing	3.09
8.	Shop-keepers	
	(b) Cart-drivers	
	(c) Tailoring	
	(d) Barber	
	(e) Carpet weaving	1.23
		100.00

ANNEXURE III

Per Capita Income Statement of Telbal Dachigam Catchment

	Sector	Individual Sector contribution	Percentage Contribution
1.	Livestock	576.82	35.00
2.	Government Sector	299.78	18.19
3.	Household industry	292.20	17.73
4.	Business	183.92	11.16
5.	Horticulture Sector	165.96	10.07
6.	Labour Sector	85.70	5.20
7.	Agricultural Sector	43.70	2.65
		1648.08	100.00

ANNEXURE IV

Energy Consumption Pattern of Telbal - Dachigam Catchment

Type of Energy Used	Annual Expenditure on Energy			Percentage of energy consumption	Per capita Monthly Household energy consumption		
	Domestically procured	Market purchased	Total		Domestically procured	Market purchased	Total
1. Firewood	Rs 307190	Rs 195412	502602	39.72%	60.95	38.77	99.72
2. Soft Coal	279618	62465	342083	27.03%	55.47	12.40	67.87
3. Agricultural Waste	270160	—	270160	21.35	53.60	—	53.60
4. Cowdung	85224	—	85220	6.73%	16.90	—	16.90
5. Electricity	—	35286	35286	2.78%	—	7.00	7.00
6. Petrol	—	18000	18000	1.43%	—	2.38	2.38
7. Kerosene Oil	—	12000	12000	0.96%	—	3.57	3.57
Total Rs.	942192	323163	1265351	100.00%	186.92	64.12	251.04

Notes: 1 Kilogram of Firewood = 2 Kilograms of Soft Coal.
1 Kilogram of Fire Wood = Kilogram of Cow Dung.

SESSION VI

OPENING ADDRESSES

WELCOME ADDRESS

PROF. SHAH MANZOOR ALAM,
Vice Chancellor, University of Kashmir

Mr. President, the Chief guest, Shri Jag Mohan, Governor of the State of Jammu and Kashmir, Dr. Raja Ramanna, Chairman, Atomic Energy Commission, distinguished Participants and Guests, Ladies and Gentlemen: It gives me immense pleasure to extend to you all a cordial welcome on behalf of the sponsors of the Seminar on the Energy Problems and Prospects of the State of Jammu and Kashmir, namely the Government of Jammu and Kashmir State, the Atomic Energy Commission and the University of Kashmir. I am particularly happy that the two key figures who have been mainly responsible for the realization of this dream into a reality namely Shri Jag Mohan, the Chancellor of the University, and Dr. Raja Ramanna, the Chairman of the Atomic Energy Commission, are with us. But for their keen personal interest and enthusiastic support, the seminar could not have been convened. The enlightened support of the Chancellor to bring the University closer to the society has been a source of strength to me in achieving this objective. It is a matter of personal gratification to me that he is with us as our Chief Guest to inaugurate the Seminar. The idea of this seminar was initiated by Dr. Raja Ramanna during the course of his convocation address at the University of Kashmir in October 1984. His interest never flagged. It is, therefore, a matter of profound pleasure for us that he is with us and because of him the Atomic Energy Commission is present here to examine the economic feasibility of locating the Atomic Power Plant in the State of Jammu and Kashmir. The response of the State Government was equally encouraging and the presence of the Deputy Chief Minister, Sir, is an ample proof of the support to the idea of the Seminar. I am very happy that despite his various pre-occupations, Thakur Sahib rushed back to Srinagar to preside over the meeting as he had some other pressing engagements elsewhere. In this regard I also must mention the name of Mr. G.D. Wadhwa, Secretary, Department of Power, who has a keen interest in the organisation of the seminar and lent full support to us. Ladies and Gentlemen, there is an acute scarcity of power in the state and the gap between demand and supply is widening. The deficit which was only 64 MW in 1979-80 increased to 127 MW by 1982-83 and is likely to rise to nearly 500 MW by the end of this decade. Let us also face this tough reality that the state has hardly any coal, oil or natural gas; hydro electricity and firewood are the only two economically feasible sources of energy available to the state. In the production of hydro power the state is climatically handicapped because of snow bound conditions in its northern half when the hydel power generation drops abruptly and literally paralyses the economy of the State. During this season the supply of power from the northern grid is also reduced significantly because of the competing demands from the neighbouring states. This acute scarcity of power in the state has led to a massive and reckless use of firewood as a source of power with disastrous effects on mountain ecology and agricultural productivity. In one of the surveys conducted by the scholars of the University of Kashmir, it has been estimated that 40 per cent of the energy requirement of the state is met by firewood. In the Naranag catchment area it has been estimated in another study by another scholar of the University that floury provides 98 per cent of the total energy consumed. In our present planning perspective the development of power is being organised sectorally and in isolation from each other. In order to cultivate an integrated approach to power development we must explore the possibility of establishing an energy development corporation in the State. This corporation will cover all sectors of energy and will attempt to coordinate and integrate their development. Shortage of power supply has caused the stagnation of the economy of the state which has consequently lagged far behind the other States of India in the field of science and technology and economic development. The object of this seminar is to evaluate critically the power situation in the State, assess its needs and suggest ways and means to minimize and if possible eliminate in the foreseeable future the scarcity of power which constitutes a major bottleneck in the progress and prosperity of the state. We must think of the various alternatives i.e. biomass, solar energy, nuclear power, geothermal energy etc. to meet the situation. In this quest for energy and progress, the University is keen to contribute its mite through research and development. We would like to explore the possibility of establishing a centre for energy studies if funds are made available by DST, Atomic Energy Commission, Department of Energy etc. This will enable the University to contribute towards an appreciation, understanding and dissolution of the energy problems of the state on a continuing basis.

This seminar is unique in the sense that the national agencies and officials of the State Government plus the University scholars have come together on a common platform to appreciate, understand and if possible suggest implementable solutions to the energy problems of the state. It symbolises the fact that for any solution of the developmental problems of the States, the national agencies concerned must be fully involved and their cooperation should be available to us. Furthermore, the universities cannot isolate themselves from these problems. Scholars in Universities must make a conscious effort to undertake research on applied problems so that usefulness of universities to the society which supports them is fully established. The seminar on the energy problems of the state is the first of the series of problem-oriented seminars which the University of Kashmir is organising in collaboration with the State Government and agencies established by the Union Government. We are planning a second seminar on problems of Regional Planning and Urban Development in this State. In this area, our Chancellor is a distinguished specialist and with his active support we should be able to organise that seminar also in collaboration with the State Government and the national agencies concerned. Before I conclude I must acknowledge my gratitude for the valuable effort and commendable work that has been done by Prof. Khera, Prof. Sharma and the team of scholars in the University to make this seminar a success. I hope our guests from outside will find their stay pleasant and enjoyable in the picturesque setting of the valley of Kashmir which is famous for its natural glory and incomparable charm and majesty. I once again extend you all a cordial welcome.

Before I sit down I must say again as I told you initially that it is a great pleasure for us that we have among us here Shri Jag Mohan, the Governor of the State and the Chancellor of the University who has been a pillar of strength to us in the development of the University on the right lines. He is here as the Chief Guest to inaugurate the seminar and I have the pleasure now to request him to kindly inaugurate the conference.

INAUGURAL ADDRESS

SHRI JAG MOHAN,
Governor, Jammu & Kashmir State

Hon'ble Deputy Chief Minister, Hon'ble Dr. Raja Ramanna, Mr. Vice-Chancellor, distinguished delegates and Participants and Organisers, Ladies and Gentlemen, it is my previlege to be with you this morning for the inauguration of this seminar. I have no doubt that with such high level participation and with such stimulating environment in Kashmir the deliberations of this seminar will be very rewarding. With our Deputy Chief Minister present I can say that the conclusion drawn by this seminar will be rigorously followed by the State Government. Seminars have little significance in my view if they are not followed up by the implementing agencies concerned. We must therefore speedily consider and implement the recommendations. Mr. Vice-Chancellor has suggested that the Central agencies will be liberal in granting help. I have no doubt that they will be so. I am deeply indebted to Dr. Raja Ramanna and other distinguished delegates for having taken the trouble to come all the way to give us the benefit of their advice. I am also deeply grateful to the departments and the ministries of the Central Government for lending their full support and also to the Atomic Energy Commission for lending full support to this seminar.

Energy: The subject of this seminar is ot crucial importance for all of us, it has always been so. Energy is the very basis of our existence. It is our 'stimulus' called *Shakti* that motive power that moves the Universe. Without this, I think, life cannot be sustained at all. Even when we talk of the food crisis, it was remarked even if world is able to produce all the food where is the energy to cook it. This indicates the multi-dimensional problem of the energy. It is not concerned only with a limited sphere but it covers, in my view, the entire gamut of human life. Most of the wars, even, as you know, have been fought to corner the source of energy. Even the international economic world order really means the distribution and the use of energy. It is the basic issue and so far as our state is concerned I think I can quote or I can narrate a few personal incidents that happened only a few days ago. I got a telegram from Jammu which says temperature touching 45°C, being roasted alive, no power for 12 hours, industry not functioning, please help. I thought I should pass this question of help to this seminar. Secondly, during the elections I came to this state and switched on the heaters in Raj Bhawan. The voltage was so low that instead of heaters heating the room it was the room that was cooling the heaters. The last week I had the privilege of going to the shrine of *Vaishnu Devi.* While I was going up it was day time and almost all the street lights were on throughout the day. Some one had apparently forgotten to switch off from Katra. But when I returned at night there were large patches of darkness all along the road. I was told that the voltage fluctuation was so heavy that whatever bulbs were put they went out of order.

Now these three incidents to my mind explain the damage that is being caused by the energy crisis, the inconvenience that is being caused by the energy crisis, and also indicate our somewhat indifferent and casual approach to the problem. We talk but we never realise the serious extent of the damage that is being done. All of us probably here must be aware that various calculations made about the damage indicate the colossal loss that our industries are suffering because of power cuts. Even a ten per cent power cut will cause a loss of 7 thousand crores of rupees per year only in the industrial field because of the heavy price of fuel, commercial fuel; the wood is being consumed, cow-dung is being consumed. The estimates show that about 150 million tons of wood is being consumed as a domestic fuel. Likewise about 1400 million tonnes of cow-dung is being used in rural India as domestic fuel. You can just imagine the damage that is being caused to the environment and in upsetting the ecological balance. Then let us come to the performance side, although we have done very well in the indigenous oil production and we have raised our indigenous capacity substantially and we were having only 37 per cent indigenous production in 1979-80 now we have about 73 per cent. Yet in other fields our performance is extremely disappointing. For instance, our utilization capacity is only 48 per cent as against the prescribed norm of 58 per cent; our transmission losses are 21 per cent as compared to 10 per cent in the developed countries and our implementation capacity should be evident from the fact that 71 per cent of our whole plan projects have been delayed and 41 of them for over three years. So if this is our implementation capacity then how do we solve the energy problem at all. Coming to our state, I think the position is far more disappointing whatever be the reasons, the transmission losses, I think, are the highest in the country, about 40 per cent and if we look around the Kashmir Valley or go around Jammu the denuded hills will themselves speak of the energy crisis, the ecological slums are really mounting every day whether you are in Jammu or in Srinagar you can see that, what is the way out? I sometimes see that quite a number of people in Jammu or Srinagar when they meet me, notwithstanding a telegram which I have quoted above, seem to have resigned themselves to the situation, power cuts, inconvenience they take it now as normal. Their attitude is:

ے رَنج سے خوگر ہو انساں تو مِٹ جاتا ہے رنج
مشکلیں اتنی پڑیں مجھ پر کہ آساں ہو گئیں

ع دَرد کا حَد سے بڑھنا ہے کہ دوا ہو جانا

So this is the actual position in our state. Now where to go from here. The basic issue is that as the Chinese say that if there is a crisis, there is also an opportunity, it is for you to see that opportunity. Mr. Vice-Chancellor mentioned that we have no oil, true we have no oil but Japan has also no oil and Middle East countries have abundance of oil. If we see the performance of Japan in the decade 1970-80 it is the only country in the world which has, with every energy crisis, increased the productivity or maintained its level. Why is it so? It is because of the technological innovation and sense of dedication and purpose. Take, on the other hand, the case of Middle-Eastern countries or even Algeria or Mexico or Venezuella. How is it that with all abundant oil their economies are stagnating; their agricultural and industrial productivity is declining? So it is not the resource, the physical resource or the natural resource, it is also your innovative capacity that matters, your sense of dedication and purpose that matters. There is no greater resource than the resource of the human mind. The inventiveness of the Indian human brain, if you can think of that; if we can work on new lines, we can develop new ideas, new technologies and a sense of purpose and dedication. Then we will be able to solve our problems. I have no doubt about it. And a seminar of this nature, of this level, has a task before it i.e. to stimulate our innovative capacity, to enable us to think on new lines, to work on new lines. Now I have four or five suggestions for this seminar to ponder over or consider. The first is the architectural aspect of energy or what I may call the climate-sensitive building style. If building is so set that the energy requirements of house or the building are met from the natural lights then a substantial problem is solved so far as the heating of the house is concerned. This is particularly relevant in areas like Ladakh where you have got more than 300 sunny days in a year. Techniques are available by which even in the coldest of the night you can keep your houses warm and if we use these techniques and if we devote our special attention to these techniques, I am sure in the cold weather both in the valley of Kashmir and in Ladakh area and other areas we can get 70 to 90 per cent of our energy requirements for purpose of heating from the sun itself. Even otherwise the solar energy is the energy of the future and there is one study in the paper that has been circulated which shows that even at the present level of research solar energy can be viable economically if it is combined with irrigation system and also with lighting system it can work and there are estimates that by the middle of the next century about one-third of the electricity will be available from photo-voltaic cells.

Unfortunately in our country we give up very easily and this is true not only of the field of solar energy where concentrated efforts need to be made because it is this sphere which will enable the Third World to out-rank the developed world. You are aware of the famous flanking Chinese General who talked about the out-ranking strategy. That is for wherever you have got impenetrable fortress, go around it. So you can always go around the industrial revolution and come to the post-industrial revolution provided you can find out a way up for harnessing solar energy. It has a great potential for us and so is biogas. You will say that due to climatic consideration Kashmir is not a suitable place for biogas. Prima facie, it is true; but why do we give up so quickly? why don't we think of new lines that can meet the problem of temperature and if we are able to do so, we will be able to convert our liabilities into assets. An estimate has been made that if biogas technology alone could be perfected in India by the turn of the century 3/4th of the energy need of rural India can be met from biogas and the remaining 1/4th from the agriculture waste and a part from the fuel wood. If this happens the consumption of kerosene will not be there for the purpose of heating your homes and cooking your food and thus the kerosene saved can work 21 million irrigation water pumps and you can just imagine what revolution in the field of agriculture production these 21 million pumps will cause because this is the kerosene which we will be otherwise consuming by the turn of the century. Then the very pattern of our existence, pattern of human settlements can change because when you have got a self-contained economy at the village level or a group of village level the rush to the metropolitan cities will decline and when that declines your energy consumption on transport and the huge road net work will also decline. So this is a great challenge for our scientists and for our thinkers. Not only in the rural areas but in the urban areas this biogas has equally great potential. I will give you one illustration. This may be partly applicable to the Srinagar Municipality and there is a paper in your seminar which somebody from Kashmir has contributed. Srinagar Municipality which shows how much petrol is being consumed, how much kerosene oil is being consumed in Srinagar and how acute the problem of defecation and sanitation is in the city. If in large slums in Delhi, for example, you go in the evening you will find large number of 'Angethees' working and all these Angethees pollute the environment and make not only the visibility poor but also cause all types of diseases like T.B. chest diseases and then in the morning it is a problem of waste. Who will collect this waste of thousands and thousands of families. You will have to employ sweepers, an army of sweepers, yet the thing will not be cleaned but if a technology can be developed to utilize this waste and converted into energy and supply to those slum dwellers they will not burn 'Angethees', they will burn that gas and there will be no problem of waste disposal. Such revolutionary potential is there provided we can make use of that. So this is another field to which enough attention has not been paid by us in my view. Third of course is the field of conservation and management. As I have said before we have our plants, thermal plants running at 48 per cent of the utilization capacity and if we are able to increase it to the standard norm which is 58 per cent there will no energy problem, there is no power problem, no power cut and this 58 per cent was once achieved by our thermal plants in the year 1976 or 1975 so there is no reason why we can't do it now and there is another study made by the inter-ministerial group which says that if you make an investment of 5100 crores of rupees in maintenance and the energy conservation measures you can save energy worth 3100 crores of rupees annually and if you want to produce equivalent amount of that energy you will have to spend 8000 crores of rupees. This should underline the need for conservation and the need for better management. There is no reason why we should not be able to increase our productivity and have more discipline and dedication to that work. Fourth field is, of course, that of nuclear energy of which Dr. Raja Ramanna is the last word and I would not dwell on it. All that I can say is that there may be a scope for development of that energy at some Central point keeping in view our requirements of security, and other requirements, and then feed it to the Central grid from which we can also benefit. The fifth and the final area in my view is the most vital, the most basic i.e., our social, economic and cultural system. At the very root of the energy crisis lies the crisis of our values, the crisis of our life style, and the crisis of our consumption pattern. If the economic and social forces operating in the nation divide the nation into very rich and very poor. In it own way that will spoil the ecological balance and if our social and cultural ethos is such then we generate a sense of indiscipline and attitude of casualness and indifference then our innovative capacity will always be undermined and we will all be wasting our time in petty quarrels and squables and we will never be able to build an energy-efficient society. Even at the international level the same thing is true as long as we have the world divided into two blocks - the block of the poor and the rich. The former will over consume the energy and the later will be left with no resources to tackle the problems of energy and the consequent problems of environment degradation and ecological balance. Both at the national and the international level we require a social and cultural revolution. But the basic and the crucial issue is who will supply the energy for the required changes, for that revolution. This is the basic issue. I am grateful to you for having listened to me and I can assure you that so far as I am concerned I will put my little weight in favour of the implementation of the recommendations which this high level seminar might make. I wish you the best of luck and best of time. Thank you very much.

KEYNOTE ADDRESS
Problems and Prospects of Power in India

DR. RAJA RAMANNA
Chairman, Atomic Energy Commission

Hon'ble Governor, Shri Jagmohan, Hon'ble Deputy Chief Minister, Shri Thakur, Mr. Vice-Chancellor, Prof. Manzoor Alam, Secretary to the Ministry of Energy, J&K State, Prof. Khera, Ladies and Gentlemen.

INTRODUCTION

We arrived here very happy yesterday knowing that we were coming to the most beautiful state in this country and we could enjoy the beautiful trees and the surroundings which the Governor referred to, and we did indeed see the tremendous changes that are taking place in Srinagar, the lake scene is cleaner, there are more trees, people look better fed and better clothed and so it was indeed a very pleasant arrival except that late in the evening I heard of this frightening air crash (of Kanishka) and this morning just before coming here I was told that my old friend Y. Nayudamma was in the plane and he must have been killed. Prof. Nayudamma used to take a lot of interest in energy planning and so you will forgive me if I make a note of sorrow in the beginning of this rather important meeting that we are holding today. Anyway as we said on another occasion when somebody passes away like this we have to merely say that the world must go on and we have to keep planning for the future of our country and the rest of mankind.

The Governor and the Vice-Chancellor were good enough to refer to the convocation that was held a few months ago in which I was made an old alumnus of the University for which I am very proud and grateful. In passing, there was mention of the energy problems of the State and the Governor, I think, was the first to suggest that we have a discussion on the subject. Both the Governor and the Vice-Chancellor have given me the credit for the idea of this Seminar but I think it was really a tripartite agreement. We thought it would be a good idea that each state needing energy discusses its problems with those who plan and those who can suggest other possibilities and alternatives. We are really here today mainly due to the enthusiasm of the Governor, because in other places nothing came out of similar suggestions. This is the first time, in this lovely lecture complex, we are going to have this discussion on the problems of the State specially with respect to energy. However, the Atomic Energy Commission does not recognize States because we believe the states have been created for linguistic, political and other purposes. But purely from the point of energy, we only recognize grids and then the discussion becomes a little easier. However, listening to the Governor and the Vice-Chancellor, particularly the Governor, his statements were very challenging, in fact they almost placed a heavy responsibility on me as to what I should say and how I should follow up the various suggestions and comments that he made. In my keynote address I will therefore follow this procedure.

I will first talk of what is known to most people as priorities for power and then about the dependence on power in the various sectors. But I want to make a special reference to an aspect which the Governor made and that is even if you have installed power why are we not getting the full benefit out of it. Where are the troubles? How can each State help in improving an efficient use of the power that is installed, power that is generated? The Governor himself gave many instances of how industry is suffering, how lights are switched on during day and there is darkness in the night, that must be the effect of the Sun on us. But I will discuss these problems at the appropriate time in my presentation this morning. My colleagues have also come prepared with all the experience that they have on nuclear power generation. But I will take a wider view of the various sources of power that are available to our different grids, and what problems they face both from the point of view of expansion and from the point of view of maintenance. Now it was already mentioned that nearly 40 per cent of India's total energy requirements are met from non-commercial sources like firewood, agricultural waste, animal power and manual labour. The remaining 60 per cent of the energy requirements is classified as commercial energy of which the contribution of electricity is estimated to be 30 per cent. The industrial and agricultural sectors which contribute most to the economy consume over 75 per cent of the total electricity generated. These sectors being the main contributors to the national economy thus depend heavily on the power sector and it is imperative that this sector be alleviated of all its problems as far as possible.

Now we shall discuss the problems that are being faced by the three options for power generation that have been chosen by the country, namely, hydel, thermal and nuclear. When I say we discuss the problems, I shall go through what the Governor partially stated: what is it that the consumer wants? He wants stability of voltage, stability of frequency. For those of us, who live in Bombay which is one city in this country where the electric supply is almost continuous, there are fluctuations in voltage and frequency but I think they are small compared to the rest of the country but large compared to other countries and that is because being in a great industrial city people realize the importance of voltage fluctuations and money. But if you leave Bombay, and go to any other city, be it Delhi, Calcutta, Bangalore, Madras or, I am sure, Srinagar too, though I have not stayed here long enough, the way lights dim down, come up, is really frightening. How can any industry prosper? An industrialist will require an assured supply of power if he wants to start an industry and he must get quick connections without going through applications and years of waiting. There must also be a system in which there is an early detection of the faults and servicing is done immediately. I do know that even if a fuse goes off on the main line sometimes it takes a day or even more in the bigger cities to rectify and I do not know what happens in the smaller cities. I would also like to talk of the problems of transmission to get electricity to the places where it is required, and the early warning systems of any cuts so that people can plan their programmes and above all of the low cost of power. Unnecessarily the man in the street should not be fleeced by high tariff rates. At the same time it must be realistic.

Planning for Power

Reliable power can be ensured by installation of an adequate generation capacity. The planning for power generation capacity is one of the most important and complex exercises which are to be undertaken for the development of the country. Given the vast dimensions of the power system the criteria for arriving at an optimum mix is hard to define especially if one takes into account every facet of the power system expansion and operation and their impact on the country's development. There are several constraints in achieving an optimal solution and very often it is found that we are forced to adopt sub-optimal options. The approach to this problem has been considered for the various power generation alternatives. They have to be evaluated taking into account various factors like finance and other resources that are available and arrive at a desirable pattern of generation which would satisfy the optimal criteria and at the same time remain practical. It has to be appreciated that for a vast country like India evolving a long-term plan for generation of electricity is not a matter of working on established techniques. It has many uncontrollable and imponderable factors governing the growth of demand for electricity on the one hand and the availability of resources on the other. This makes the problem of arriving at a precise solution even more complex and they are truly Indian problems. Considering the problem of finances, supposing we had an infinite amount of money, we could go on installing more generation capacity in the hope that if you have excess capacity you have plenty of power for industrial development. The other problems are how to maintain it, make it efficient, etc. But the fact remains that we have many problems peculiar to India. The Central Government was very courageous when it dealt with the agricultural problem, the food problem. In fact they exempted agricultural income from income tax, subsidized various inputs to a great extent and that is why the food problem has indeed been solved and we talk of exporting food and I think we have exported food. But people must realize that our foodgrains are expensive, probably the most expensive in the world. I have not verified this but I think it is a fact that we have subsidized them so much that we produce grains more expensive than in any other part of the world. I have nothing against it because the moment you produce food you have food in the country at least. But what is the next priority, food was the right priority. I mean a country like India which used to face famines has become a food surplus country to a certain extent.

The next priority must obviously be for power. I think everybody in the country who has anything to do with industrial planning will recognize that power comes next. But how do we get the money and see that it can be subsidized in the first instance and then generate more money by efficient utilization of the installed capacity. How much subsidy is adequate, how much extra importance has to be given to this programme are matters for the Planning Commission. The States also must give proper weightage to the available resources and plan accordingly. For example, in your State you must bring the necessary pressure for the development of hydel power within the State because that is the resource that you have. You have no gas, you have no coal but you have plenty of perennial water. But there are some problems about which I shall talk when I say something more about hydel power.

Options for Power Generation

As you all know our country has chosen three options for the generation of electricity, namely, hydel, thermal and nuclear. The present installed capacities of these modes of power generation are 24,460 MWe of thermal, 14,338 MWe of hydel and about 1,000 MWe of nuclear. Commercial feasibility of the other sources of energy such as solar and bio-gas is yet to be established. Now I say this because the Governor laid great stress on bio-gas and solar but I am talking of power as an industrial product and its cost from various modes of power generation has to be examined. Solar power is still expensive, but it may be necessary at certain locations. I have seen this in Ladakh when I was connected with defence, how useful it was in far off areas, though by commercial calculation it was a very expensive form of power but it was needed as there was no other source of power. You must have some source of power irrespective of the cost which reminds me of the famous statement of Dr. Bhabha that "No power is more expensive than no power". The power supply position is not satisfactory in many parts of the country at present and one of the main reasons for the situation is the inadequacy of the generation capacity and power transmission systems. Hence the immediate objective would be to correct the inadequacies of working, increasing the generation capacity and at the same time streamlining the working of the running power plants and transmission and distribution infrastructure. These short and medium-term objectives may not be the optimum solutions in the long run because of the rapid increase in the demand of power and development of new technologies which alter the economics of the various options. Now with this in mind, we can discuss the performance and problems which face the various power generation options.

Hydel Power

Given the technology available at present hydro power is the only available renewable source of energy for bulk electricity production and it has been accepted that it should be exploited to the maximum extent possible. The running cost of a hydel power plant is lower compared to any other option and it has the unique characteristic of being free of pollution. It is therefore prudent to develop hydro-electric power generation facilities wherever possible. India has an estimated 75,000 MWe of hydro-electric potential of which only 14,340 MWe has been exploited, and plans are underway to increase this capacity to 40,000 MWe in the 1990s. It should be understood that hydel plants can be set up only at locations which have abundant water sources. Further, this type of plant involves very long gestation periods and high construction costs. Often hydro projects are linked with irrigation thus placing restraints on power generation due to conflicting priorities. A major problem in tapping hydro-electric resources is paucity of funds. I again underline this because this is the situation as it exists. Bulk of the untapped hydro-electric resources are concentrated in States and Union Territories which are relatively less developed and do not have the financial resources to take up large hydro-electric projects. Even the States which are financially better off have opted for other options which yield benefits in a shorter time-frame. On account of these problems, the Central Government has employed a judicious strategy in planning the hydro-electric facilities in the central sector in different regions of the country for the overall benefit of the entire region. This step has met with limited success due to the fact that States with an enormous potential of hydro-electric power have not been willing to part with their resources. This is a very unfortunate thing that happens in our present structure. Environmental and ecological considerations have further compounded the problem towards the development of hydel power sources. Now I know the previous Prime Minister was responsible for the non-development of hydro potential in the Silent Valley of Kerala and I think she had very good reasons for doing this. However, one has to consider how many are the ecological considerations and how many are the industrial developmental considerations. Together with all this, uncertain monsoons have also contributed in a large measure to the risk of placing too much faith in hydel plants. I know States where they had 80 per cent cuts on industrial use of electricity because the rains failed. Moreover, problems of silting have been noticed in many projects which has led to derated capacities. Some of these problems can be overcome to some degree by afforestation and desilting programmes. In fact, I would like the ecologists here to consider the fact that artificially grown forests solve ecological problems better than natural forests. I am told that our forests, including natural forests, have deteriorated very much and I know that in European countries they grow artificial forests in a very short period of time which counteract ecological disturbances. I have a feeling (I am a little arrogant and cocky on this) that it should be quite easy to grow good forests in less than twenty-five years or perhaps even less, and that is because we had a bare hill in Trombay with nothing on it and the moment a green leaf was seen somebody used to pounce on it, either destroy it for purely vindictive reasons or take the piece of stick for making his chappati. But we just stopped the cutting of trees, planted a few trees as advised by our horticultural experts land and right in the middle of Bombay we have one of the finest artificial forests. It certainly adds to the beauty and I hope it does change the climatic conditions in some minor way. The rest of Bombay city is denuding forests and throwing poison all over in the form of SO_2 and NO_2 and so forth. A fact that we must bear in mind, particularly in the State of J&K, is that artificial afforestation is a possibility and this can be borne in mind when you are thinking of hydro projects which is your very big resource.

Thermal Power

It has hence been accepted that hydel sources alone cannot meet the growing demands of electricity because of various reasons already given. In all the regions of the country a judicious mix of thermal, nuclear and hydel plants would have to be deployed depending on the resources available at various locations. The Governor again was good enough to point out that in some central place we should provide the grids with a judicious mix so that no part of the country suffers because there were no rains, no coal or things like that. Thermal power plants based on steam turbine technology and coal as fuel have been an important source for power production in India. They account for more than 50 per cent of the total installed capacity and contribute a corresponding share to the total electric power generation in India. As a deliberate policy measure, use of oil and gas for power generation has not been encouraged, from the time of initiation of development plans except in situations where it is considered advantageous. This is because of the limited resources of oil and gas available in the country and the burden that import of oil imposes on the balance of payments. India has sizeable resources of coal. According to an assessment the total coal resources in India are of the order of 112,000 million tonnes and out of this about 88,000 million tonnes are of the non-cooking variety. Considering the fact that there is a large gap between demand and supply of electricity in our country, the limited hydro reserves available and that the nuclear option is just coming out of the developmental stage, thermal power has been the most important electric power source in the country. Now one of the main advantages of the thermal plants is their low gestation period compared to hydel projects and a lower capital cost. I believe thermal plants can be put up in four to five years' time perhapless, but if you add the time taken for the coal sources to be developed at the same time, then the gestation period becomes very long. You will see that later on there is going to be a shortage of coal of the right type. Though the performance of thermal plants some time ago was not satisfactory it has improved substantially during the recent years. The plant load factor during 1984-85 was of the order of 50 per cent. Though not considered good by itself this load factor is much better than in the previous year. However, the economics of the thermal plants is largely dependent on their location. Thermal power plants located at pit-heads, or near the mines, have the advantage of a continuous supply of coal without enforcing much burden on the transportation system. But this strategy may not always be beneficial considering the fact that transmission lines for bulk transmission of power over long distances are not readily available in our country.

This has forced us to set up thermal plants near load centres but away from pit-heads which has eventually caused considerable strain on the transportation system, hence affecting power generation. There are other important considerations like water supply for condenser cooling and the nature of the grid in deciding the capacity and location of these plants. Thermal plants in the past as well as the present have been facing problems with regards to the performance of equipment, both imported and indigenous, which has affected their performance and cost-effectiveness. There is an urgent need for attaining self-sufficiency and self-reliance, in the manufacture of high grade equipment adhering to stringent technical specifications, which can ensure reliable power generation over a long period. I think I can also say that maintenance requires the highest priority and maintenance training means good training and good training automatically implies selection of good people, able people, to look after power stations. Performance of our thermal plants can be substantially improved if indigenous manufactures bestow more attention on manufacturing practices and employ strict quality controls. Of course a sentence like this means that it does not exist to the extent it should. The situation has no doubt improved over the years, but it is still far from satisfactory. Now that is a typical Government statement coming from a Secretary to the Government (I am just covering myself). With the amount of incentives now offered to the manufacturers, we hope more developmental efforts would be made to improve the quality of equipment. Poor quality and high ash content of coal have also been responsible for sub-optimum performance of thermal plants and this is a problem requiring closer attention of coal producing authorities. I hope the Governor will protect me certainly to the extent he can, if I say that our coal production industry is in a very unfortunate condition and has been for a long time. In fact, in the days when I used to be in Delhi there was frequent mention of Coal Mafia, but may be things have improved now. Thermal stations have also contributed substantially to pollution. The Governor was formerly looking after the Capital and I think our Capital is a standing example of disgrace for polluting the air from a coal-fired station. Due to the high ash content of the coal available in India, thermal plants have become a serious health hazard. A thousand megawatt thermal station burning coal with 40 per cent ash content will produce nearly 36 million tones of ash during its operating life of 25 years. Many people will be surprised to learn that these hills of ash which lie beside these coal stations are also sources of nuclear radiation, though it has got nothing to do with nuclear power generation. Studies by the Bhabha Atomic Research Centre have shown that the gamma activity in residual coal ash due to the presence of elements like radium-226 is higher than that from an operating nuclear power station. Combustion of coal results in pollution caused by the release of oxides of sulphur, nitrogen and carbon. For example, a thousand megawatt thermal plant will discharge 40 to 80 tonnes of sulphur dioxide per day which is an irritant for the respiratory system and also leads to acid rain which can cuase considerable environmental damage. Decanted water from wash handling system also contributes to pollution. To overcome these problems installation of expensive equipment has become mandatory

major factors affecting the performance and location of thermal plants. Apart from this, the problem relating to equipment breakdown on the conventional side is common to both the nuclear and thermal plants. The working group on the energy policy 1979 had made a recommendation that owing to the growing demand for electricity and the limited options available within the country, it is appropriate to undertake a nuclear power programme on a larger scale and enlarge the technology, design and manufacturing capability. It has recommended that measures have to be taken to accelerate the growth of nuclear power in the country. Consequently, the Department of Atomic Energy studied the feasibility of an accelerated programme in the nuclear power production and set a target of achieving a capacity of 10,000 MWe by the end of this century. It should be appreciated that the design of nuclear power plants of PHWR type with a capacity of 235 MWe is now completely standardized and there is a very high probability of the forthcoming units operating with a very high degree of efficiency. This has been proved by the performance of the first unit at Kalpakkam. The main aspects that are to be borne in mind while setting up nuclear power stations are considerations like seismic criterion and transportation of spent fuel and waste. In a nuclear power station, transportation of fresh fuel is not very difficult but spent fuel is radio-active and has to be treated very carefully. Further, nuclear power stations should not be subjected to large fluctuations in power production. The fluctuations in demand should be taken care of by the conventional stations. These aspects have to be carefully studied before arriving at a decision on a site for a nuclear plant.

Work has been taken up on the development of Fast Breeder Reactor technology which would save the way for meeting the country's demand of power in the long run. Fortunately the atomic energy programme has been one in which we have tried to look a little ahead, though we were not quite ready for it, so that we are not kept wanting when the time comes and that is why we started work on the breeder technology quite some time ago and the Fast Breeder Test Reactor (FBTR) at Kalpakkam is practically complete and should go into operation in August/September this year (1985). So it is a rather important year for us, when three reactors will be going into operation – Dhruva in Bombay, FBTR in Kalpakkam and MAPS-II unit at Kalpakkam – all well before the end of the year. Now I did mention fast reactors and I have to look a little bit into the future. We have a tendency to look a little too much into the future and not to the immediate requirements which is the guidance given to us by the Governor. But one good thing about Indian thinking is that we do think a little further than many other countries because he did mention how countries with a lot of oil are not getting the benefit of the oil whereas countries without oil get all the benefits. We do not want that to happen. We have a large amount of thorium in the country and we would like to use that by converting that into uranium-233 in a breeder reactor. At the moment the economics does not seem good but it has been my experience that the economics of nothing in the beginning seems good. It has to be developed and at a later stage a decision has to be taken. But the way the fast reactors are working, the breeder reactors have been working in France, and I think the Soviet Union also, seems to indicate that they are a very good alternative source of power for the future. In the process, the technological development of the country progresses very much because these reactors are cooled by liquid sodium and it is an entirely new type of technology. It is not the purpose of this keynote address to tell you all about breeder reactors, but I just end on that issue by stating that the breeder reactors are those in which more fuel is produced than is burnt, that is, as the reactor is working, more fuel is being produced outside than what is being burnt inside. So from a somewhat romantic point of view the fuel problem is solved but nothing is solved to that extent but in reality fast breeders ease up the fuel problem to a very great extent. This remarkable system is possible only with nuclear development. I talked about this only to give you a glimpse of the future but let me come down to earth again. After all we came here to solve problems and we must first identify what the problems are and then my colleagues and others will take over.

Problems in Power Sector

I think the biggest problem in the power sector which I would like to refer to and then sit down is that related to transmission and distribution. Though I am not an electrical engineer, I would say that inadequacy of the transmission and distribution system is a major bottleneck for industrial development in all the States. It is a recognized fact that one of the most important and crucial areas of the power sector is planning, implementation and operation of an effective and efficient transmission and distribution network. Yet this has not received adequate attention resulting in a very weak network. A lot of public attention is drawn towards the poor performance of power plants while little or no attention is focussed on the poor state of our transmission and distribution system, e.g. about 20 per cent of the outages in some of our power stations have been due to poor grid behaviour. The capacity utilization of a power station also heavily depends on the grid. The performance of the grid thus affects the efficient and economic operation of a power plant and therefore there is an urgent need for a strong transmission and distribution network. The main reason for the poor state of affairs is a lack of coordination between the development of generation capacity and transmission and distribution systems which in most cases results in under-utilization of the generating capacities. For a vast country like India, where power resources are not evenly distributed, transmission and distribution systems have to be developed in a highly systematic manner at the national level which can be broken to regional levels with the main objective of ensuring a stable and strong grid and at the same time taking care that power resources are evenly shared. *Ad-hoc* additions to the transmission and distribution systems at the State level or at regional level alone cannot help in achieving this objective. Another factor contributing to the lack of development of an efficient distribution and transmission system is the paucity of funds for the purpose. Payments to the power generating organizations have not been regular leading to financial problems. In fact at the end of May, 1985, arrears outstanding from various State electricity boards to nuclear power stations alone were about Rs.90 crores.

Now if you do not get that money then obviously it cannot be ploughed in. Further loss on account of theft of power which is of the order of 15 to 20 per cent (these numbers were given to me. I have not verified the theft but it gives you a feeling regarding the alarming situation in the power distribution system which has been a contributing factor to the poor financial position). Lack of coordination between various States has also crippled the development of an efficient transmission and distribution network. Utilization of funds provided by the Centre for intra-and inter-State bulk transfer of power has not been satisfactory. Inter-State and inter-regional transfer of power has a number of bottlenecks and a lack of consensus between concerned States has been retarding the establishment of an optimum system of power distribution. There is an immediate need to integrate the power generation plans with transmission and distribution network plans if one has to witness an efficient utilization of generation capacities created. A national plan should be evolved perhaps with the use of a computerized model for the design of an efficient transmission and distribution network. A positive step has been taken in going for a 400 KV line for bulk transfer of power. This would however solve the problem only to a limited extent because the main drawback is a lack of discipline on the part of State electricity boards for transfer of power from one division to the other. The Government has therefore, set up a committee to consider the installation of a high voltage d.c. link between super thermal stations and the power deficit States. This would ensure allocation of specific blocks of power to each State. However, the economics of such a system and import of equipment to begin with will be the main considerations which are presently under evaluation. The d.c. transmission may help in solving a sociological problem – it is rather a strange way of solving it.

Conclusion

In conclusion the discussion so far has revealed that the power scene in the country has various problems. All the options for power generation – hydel, thermal and nuclear – have problems, some of which are common while the others are unique to the specific mode. Further, it is seen that the transmission and the distribution systems also need considerable improvement and there is a need for a grid at the national level. Hence our national power plans should strike a suitable balance of adopting an optimum mix of various options keeping in view the long-term perspective. One cannot dispute the fact that we need a mix of all the three options open to us at present to augment our power capacity. Hydel sources, as already mentioned, should be exploited as far as possible especially in States like J&K, and other areas in the Himalayan ranges, which have perennial sources of water. This will solve most of the power problems in these regions. Today thermal and nuclear power facilities face similar problems with respect to the conventional power generation equipment. Thermal plants have pollution problems while nuclear plants have the task of safeguarding from radiation exposure but these are fully under control. While thermal plants are incorporating equipment to reduce the effects of the release of toxic gases, nuclear power plants already have one of the most sophisticated safety systems to safeguard against the dangers of the release of radiation. Taking into account all the cost components, nuclear power plants compete favourably with thermal plants even for units set up at the pit-heads.

Now this is an interesting controversy because if you can produce nuclear power as cheap as coal power at the pit-head, well somebody will say, why don't you set up nuclear power stations even on coal mines. But then what will you do with the coal. Are you going to send coal all the way somewhere else and take all the cost of transporting that coal? Then obviously the situation comes about that it is better to put nuclear plants away from coalmines and use the coal where it is. But there is much discussion in the country that since the cost of nuclear power is almost the same as that of coal-fired stations at pit-head you could put it anywhere you want, but this is something which is an interesting factor which we should take into account in our discussion later on. In the near future the thermal plants would continue to produce a major fraction of the total power in the country. In the long run we shall witness nuclear plants taking over the burden significantly as our fossil fuel reserves get depleted (and expensive) while the demand for power increases rapidly. Well your Excellency, the purpose of my keynote was to expose the problems to you not necessarily at a State level but at a national level with fall-outs to what the States should be doing. None of the problems has an easy solution, but I think you made a very important point towards the end of your introduction and that is the sociological problem. Attitudes have to change – that it is my property and I will not let the water go to the next State – seems to be a very old-fashioned attitude to property, whereas if power is available to both States and both benefit by the industry that is the answer. Actually in Europe one country is proud of sending so much to this country, so much to that country and I have a feeling in spite of all the differences between the Soviet Union and West today, they exchange power between one another quite frequently. I have not checked this but I think this is true. I think they sell power to one another. Now that attitude must come about and in that case a great many of these things will follow. There are some aspects which I may have started with which many of you may not agree with but what is the function of a keynote address if it is not stating some strong personal views and also getting the benefit of experience and discussion with our colleagues? I thank you for listening to me for a long presentation and I hope it will be useful and we will follow it up.

CONCLUDING ADDRESSES

VALEDICTORY ADDRESS

PROF. MAHESHWAR DAYAL
Secretary, Department of Non-Conventional Energy Resources, Government of India

Honourable Chief Minister, Prof. Manzoor Alam, Prof. Khera, Mr. Wadhwa, Distinguished Ladies and Gentlemen, and participants to the seminar.

I am indeed very very happy to be here this afternoon to participate in your valedictory function. It is always a pleasure to come to Kashmir. It is a beautiful State. It is endowed with such gifts that Nature has bestowed and it should, infact be our endeavour to see what we can do to help foster economic development of the State while at the same time preserving the natural beauty and environment which has been given to us in such abundance in this part of our country by Nature itself. It is this special characteristic of Jammu & Kashmir State which I think lends particular point to the type of energy development that we could consider where we can have a system of meeting the needs of our people in a rapid manner but at the same time in a manner that preserves the environment and the ecological balance and preserves the cleanliness of its atmosphere and of the rivers and the lakes. Now we have heard, I am sure during the course of the seminar as it is also been summarised by Shri Wadhwa and others, discussions on various forms of energy. So I will not go into this entire matter of different forms possible but I would like in my address to concentrate on the forms with which I am dealing at present, i.e., the non-conventional energy forms. These are in many ways very much synonymous what I thought I mention earlier as possible the best objective of the energy development in the State, indeed in the whole country in my opinion, that is the development with environmental preservation. This State, I think is particularly well endowed with resources of new and renewable sources of energy. In many parts there is a great deal of solar energy available and there is a lot of wind energy available. I had the pleasure of visiting a few months ago the entire region of Ladakh where most of these energy needs can be supplied from just these two sources alone, i.e., solar and wind energy, including power. There is a lot of bio-energy available in both Jammu region and in the valley and if we look at the consumption patterns of energy we find that in fact that a great deal of energy needed today for domestic purposes is in the form of heat. This is supplied by biological sources such as wood, wood wastes, agricultural residues, animal waste and so on and at the same time we need electric power and for this purpose we can use all these sources, I have mentioned, and also the hydropotential of J & K State which abounds in small streams, rivers as well as larger flowing systems of hydro. But it is entirely possible to meet a great deal of electric power needs by installing smaller hydropower stations which are known as mini and micro hydel power stations which are of short gestation period and which gives you the power required at the local level in a quick pace of time. I therefore, feel there is a tremendous scope here for what we call the non-conventional systems and I would like to recall briefly that in the last year, thanks to the support and participation of many of our colleagues throughout India, several of them sitting here who have actively helped and contributed in this venture. I see here Prof. K.L. Chopra of the I.I.T., New Delhi, Prof. Thaper from the University of Roorkee and a number of others who have been instrumental in leading this revolution which is beginning to take shape in our country.

You will be interested to know that during the last year the number of solar systems that we have installed for water heating, crop drying, cooling, timber kilns etc., in the country were such as to be equivalent to the generation of 200 million units of thermal energy per year. The number of biogas plants installed during one year were more than 181,000 which is equivalent to a saving of something like 7 lakh tonnes of wood per year and also fertilizer worth Rs. 30 crores per year. This was done alone with an investment of only about 47-48 crores so this is another very big programme. The third programme which has made a very big headway in terms of actual implementation is the programme for the high efficiency wood stoves which tend to cut down the consumption of wood by a factor of two with a very small investment because these high efficiency stoves cost only between Rs.40-90 per piece and in a very quick pace of time by installing it we can suddenly drop the consumption of wood which is one our objectives. In this one year we have installed more than 80,000 of these wood stoves throughout the country which is equivalent to a saving of wood consumption of 6 lakh tonnes per year which is about 24 crores of annual return from this one source alone. Now in addition to that there are number of other devices. Some 160 remote villages have been electrified using solar photovoltaic systems and more than 1000 wind pumps are operating and 101 community biogas plants have started functioning. Systems for getting energy from sewage and urban waste have started functioning in cities and farms and so on. Now at the other end of the spectrum a new programme which started and is giving very good results is the energy plantation of fast growing varieties of fuel wood and also new designs of hydroplants of small sizes and newer methods of using biogas, i.e., gasification techniques which can provide electric power of very substantial quantities from agricultural residues as well as wood wastes and wood. I thought that apart from discussing some of these I would like to show you some examples of some of the things which have been installed. It is always nicer to see some pictures than to hear from the lecturer.

Note: The following are some of the important slides then shown by Prof. Maheshwar Dayal to the audience:

1. One of the machines that is beginning to be installed in substantial numbers and which is meant for pelletizing agro-waste. The machine produces a sort of fuel brickets in various sizes which can be used in furnaces, *angeethies* etc. as fuel.
2. Picture of integrated energy system in village 'Masoodpur' based on biogas utilizing human waste and cattle waste for producing gas.
3. Biogas generation designed for low temperature region specially suitable for J & K State. A solar canopy uses 'greenhouse effect' to trap solar energy and thereby create heated atmosphere for the entire biogas plant. With this arrangement the generation of biogas in winter did not decline as it used to be earlier. Further a by-product slurry after being dried is used as fertilizer.

4. Wind pump integrated system: based on biogas and wind energy for the distribution of water to the surrounding area; the pump is working in Masoodpur where the water level is very low. As a first step electric pump operating on biogas lifts the water from 60-70 metres below the ground to the ground level and then the wind pump lifts water from the ground level to overhead tanks for the supply of drinking water.
5. Community Centre at Masoodpur highlighting the use of solar energy through photovoltaic pannel for the purpose of lighting, radio and TV which has proved a big draw and source of excitement for the villagers around.
6. Lamp run on biogas; another illustration for the use of biogas for the following direct applications: (i) lighting, (ii) cooking, and (iii) electricity production.
7. Wind pump: popular in high wind velocity areas like coastal areas of Orissa; it has the capacity of drawing 1 lakh litres/day and is proving very economical.
8. Another wind pump being used in a village near Bangalore to supply water to a storage tank for drinking purposes.
9. Large-scale institutional biogas plant set up at a dairy farm in Madhya Pradesh (Indore); useful where you have large number of head of cattle, sheep horses etc. for producing large amount of biogas and fertilizer. The biogas can be used for heating, cooking, producing electricity etc.
10. Example of the use of solar energy in a village in U.P. The village had a quantum jump in technology in the sense that it was unelectrified and it was provided with solar powered colour TV and the people were watching TV programmes beamed to it from a satellite even at the time when entire surrounding area including Delhi had a total black-out of power. The village is around 45 km from Delhi.
11. Another example of use of solar energy in Chejja village which never had any electric power. Solar energy is providing power for lighting and operating a water pump.
12. Manufacture of high purity silicon for solar cells being developed at the Indian Institute of Science, Bangalore, and the space-quality solar cells for use in satellites.
13. Example of a village in Tamil Nadu which uses non-conventional energy sources and all its energy requirements are met from sources available at the village. It is

using solar energy water pumps and street lighting from solar system etc.

14. Pump run on biogas for minor irrigation purposes and drinking water purposes in a village in Andhra Pradesh.
15. Newer development which is coming up at the same village in Andhra Pradesh: to provide solar energy through the thermal route for providing fairly large amount of power by using concentrating type parabolic mirrors. The paraboloids focus the heat of the sun to heat a fluid which then generates power. This type of development in some cases is proving economical especially in those areas which are remote from transmission network.
16. New Development of Sewage farm in Delhi where digesters are converting part of sewage into methane which is a fuel and is being used for energy purposes. Some 2000 connections are already given for domestic cooking. This is an expanding energy source.
17. New Development of gasifiers: to provide 30-40 KW of electric power at village level. The gasifiers can be run on wide variety of agricultural residues like corn, groundnut husks, rice residue, wheat residue, saw dust, wood wastes and so on. These gasifiers can meet all the requirement of irrigation pumping, lighting and so on.
18. One example of Energy Plantation: it is a mixture of high-yielding and fast growing varieties of trees and shrubs giving good calorific value per hectare per year even on not a good quality soil. It can be used for producing electricity when coupled with biomass. It has very interesting potential for providing high amount of employment to people; it can provide biomass cover on what could be arid land, can provide fodder for cattle, fuel wood for the poor and others. It can be used for electric power through gasifier.
19. Solar thermal system for water heating. This system is operating in Bhopal to supply hot water to dairy farms and is very economical. It is supplying 60,000 litres of hot water which amounts to the saving of Rs. 4 lakhs per year in furnace bill. There is hardly any running cost. One can have a back-up system for periods when there is no sun say during rains etc. Back-up system is always small consumer as compared to the main system.

20. Solar powered water heating system especially useful for hospitals and hotels. These have been installed in Lodhi hotel, Janpath, medical institutes etc.
21. Solar distillation unit for producing distilled water for hospitals and batteries, motor garages etc.
22. Community type of solar cooker for community cooking; installed in a school for preparing meals for school children.
23. High efficiency *Chullah* which is an excellent example of the application of the understanding of science to every day problem. In this chullah the consumption of wood is reduced by a factor of two; it is smokeless and therefore saves the women folk from lot of problems especially of health. It saves their time in cooking and the time saved can be converted into some income for the family. Time is saved not from cooking alone but in the sense of collecting wood required for the chullah and also time is saved in cleaning utensils which is much easier now because it is smokeless.
24. Other examples of chullahs with different models including portable one etc. We have a programme to give fullest support for the dissemination of these improved stoves, large-scale support for training purposes; subsidies for the hardware for construction of chullahs. This has great potentials in my view also in the Jammu & Kashmir State.
25. This is an example of smokeless village. We have a programme for making villages smokeless and under this programme 5,000 villages have already become smokeless.

At the end I would like to mention that the Jammu & Kashmir region appears to have very favourable conditions for the applications of all these technologies which are based on local resources, which can provide energy in the form you wanted efficiently whether it is electricity or heat energy and which can provide this in a quick way. One of our problems is that we start projects that take 10-15 years to materialise. The output is not available for improving production processes or the standard of living in quick manner. But here many of these technologies, however, have the potential of giving you output quickly, generating incomes, improving income levels and conditions of life in a very quick way. So I would say that it is very important, in my view, that very serious attention be given to the development of non-conventional energy forms in the Jammu & Kashmir State and I would like to assure everybody present here that from the point of view of the Central Government we shall give fullest possible assistance and support to any of these efforts. Many devices and projects have already started being deployed in the State. In Ladakh we have already started putting solar air heating systems for the comfort of the people living there. Solar cookers have been put in. Solar foot-hold type pumps have been put in many places and we are ready to do more. But what I would also request is that in the J & K State proper organisational structure should be set up to enable this thing to go in a much faster manner. The programmes have come to a stage that it is desirable, just as the Government of India at central level have set up a separate department for non-conventional energy sources it is essential to have some agency, authority or a department or a body specifically dealing with this area. Many of the States have now set up what is called energy development agencies, authorities or even corporations to deal with the development of these new sources of energy. I think that it would be very useful thing, if at least some body is set-up of this type with adequate powers and adequate financial resources and adequate qualified personnel so that the programme can be accelerated in this state and also the central assistance which we are giving in this regard can be properly channeled, increased and can be best utilized. With these words I would like to conclude my address and congratulate the University of Kashmir for organising this seminar and I would like to express my special thanks to the Hon'ble Chief Minister for the encouragement which he has so kindly given to this problem in this area and with his presence here encouraging us in our discussion of this very vital problem that is so crucial to the development of the economy of the Jammu & Kashmir State. Thank you very much.

CHIEF MINISTER'S ADDRESS

G.M. SHAH
Former Chief Minister,
Jammu and Kashmir State

Address of the Chief Minister at the Valedictory Function

Prof. Dayal, Prof. Manzoor Alam and friends: I am thankful to the University of Kashmir for having organised a National Seminar on Energy Problems and Prospects in Jammu and Kashmir State in collaboration with the department of Atomic Energy, Government of India and Power Development Department of the State Government. A number of papers have been read during three days of your deliberations and I am told that discussions on these papers have been lively and fruitful. The deliberations of the seminar have been enriched by the presence and participation of eminent scholars and experts in the country in the area of energy. This has provided a unique opportunity for the interaction between these experts and the researchers from the University and the planners in the Government. I can assure you that the recommendations of this Seminar will receive our due consideration and I accept the suggestion of Dr. Dayal that we must have an energy development authority in the State which we will set up immediately because we are very much interested in energy. We have no energy resources here by which we can even supply the needs of the domestic requirements of our population. We in this part of our country are confronted with an acute energy crisis which in turn has retarded the speedy transformation of State economy besides magnifying the various problems of all types of consumers. The generation of power in the State has not kept pace with the growing energy requirements as a consequence of which the living standards in the State remained much below the national average. It is expected that with the completion of Salal Hydel project and other on-going projects in the State the power supply will increase. The supply-demand gap is not likely to be bridged to any significant extent. The Seventh Five Year Plan of the State with an approved outlay of 1,500 crores has a meagre outlay of 276 and half crores i.e 18.5 per cent on the development of the hydel power in the State. I understand that at the end of Seventh Five Year Plan a short fall of 33 per cent with respect to our power needs shall still remain which no doubt is an alarming position. My power engineers are brooding on this problem. I think Prof. Dayal has come to your rescue. He wants to take biogas for irrigation purposes in the State and I think it can be done. It is indeed unfortunate that in spite of there being identified hydel potential of about 10,000 MW in the State, we have been able to harness only two per cent so far which will go up to 4 per cent by the end of Seventh Five Year Plan. Given the resources we could have not only become self-sufficient in power but also exported it to other States and more so we could give 70,000 crores of income to India by using water potential of Chinab, Jehlum and Tawi. I think Prof. Dayal will take note of it and recommend to the Government of India that in the whole plan first preference should be given to the power potential in Kashmir. Needless to mention that the availability of cheap power could prove a catalyst for complete socio-economic transformation of State. It is gratifying to know that the seminar has taken a note of the inadequate share of the State from the projects in the Central sector and we shall be approaching the Central Government to effect the equitable distribution among the States. This may ease the energy problem to some extent temporarily. We have also taken effective measures to minimize the transmission losses and these have yielded good dividends. With a further reduction in transmission losses during the coming years the situation should improve. In our State the vast majority of the population meets their requirements of energy from non-conventional sources and I hope that the steps taken by the State government for the nationalisation of forests and adoption of social forestry programmes the problem of denudation of forests will be checked and the economy will be benefited through quality team work and other forest-based economic entities. In view of the diverse and varied sources of energy the solution of the problem lies in effective coordination of the research and development activities in the different sectors. The recommendation of the seminar for the establishment of State Council of Science and Technology to achieve this objective is well spoken and will establish it. I am also happy to note that in view of the urgent need to undertake and promote research in different facets of energy sector the seminar has recommended the establishment of a centre for energy studies in the University of Kashmir. This will indeed be a step in the right direction and I will request the Vice-Chancellor to take action in this regard in the University. My Government will take earliest possible steps for its establishment; no, he will establish it and I will help him. In this gathering I wish to recall the challenge I have thrown to our energy experts way back in 1980 when I took over as power minister. In a similar seminar held in the Regional Engineering College, Srinagar, I told them to find and provide for our people an air-conditioned shirt and trousers, which would be the only real answer to their problems. I did not ask for the moon which too is no longer unknown to man. Gentlemen, this is the age of miracles and men's landing on the moon has become a part of history, we are talking of star wars and endeavouring to unravel the mysteries of universe. Is it not possible to find an answer to this challenge? Of course, it is possible but I wish to convey them that our experts should initiate a continuous research and find a solution to the problem, I posed; I reiterate my challenge with a fond hope that our scientists will rise to the occasion and bend all their energies to meet it. Before I close I once again thank the University of Kashmir and the Department of Atomic Energy, Government of India, and all the distinguished participants for giving to the State Government the benefit of their well considered recommendations which will receive our earnest consideration. Thank you very much.

PRESIDENTIAL ADDRESS

PROF. SHAH MANZOOR ALAM,
Vice Chancellor, University of Kashmir

Hon'ble Chief Minister, Shri G.M. Shah, Prof. Maheshwar Dayal, distinguished guests, ladies and gentlemen, I must at the outset express my personal gratitude to the Hon'ble Chief Minister for his very encouraging response to the recommendations of the seminar. This has been the most encouraging and heartening part of the entire seminar for we have here the highest authority of the State telling us that the major recommendations of the seminar with regard to the establishment of the energy development corporation is accepted and when Mr. Shah says it, it is implemented. Further with regard to the State Council of Science and Technology he assures that this will also be established; that is basic step in the right direction and I must congratulate him for having taken a decision on that and I am extremely grateful to you, Sir, for asking me to set-up the centre for energy studies in the University of Kashmir and I can assure you that I will implement it and will come to university council for its concurrence in the next meeting of the Council. But we shall implement the decision and I am sure now that you have said it, I will do it. I would like to express my personal appreciation and thanks to Prof. Maheshwar Dayal for having very graciously responded to my personal invitation. In fact there were some doubts whether Prof. Maheshwar Dayal will be here with us or not but somehow we managed and he didn't need much of persuasions to come here because I found that he was very keen to come and interact with us and I am really very happy that he is here with us and has given a most illuminating talk. Let me assure him that we are as eager as he is to implement his ideas. The University of Kashmir is on its way to innovative thinking and in this it does not lag behind any other Indian University. So you will find now the era is coming when some of the major innovations will really emerge from the University of Kashmir. We take up that challenge which he has thrown before us now that we are going to have the centre for energy studies we hope that we shall take up the job very earnestly and I am sure, as promised by Prof. Maheshwar Dayal, that Central Government will make adequate funds available for the researches in the non-conventional sources of energy. For us in the University of Kashmir the seminar has been a most rewarding experience because this has provided my teachers and research scholars with an opportunity to interact with scholars of eminence in the energy area of specialisation from all over the country. This is also an area where we find that the University Scholars have really benefited when they have interacted with administrators and the planners, energy planners of the State because we are no longer in the theoretical area of thought but in the practical area of application of energy planning and development. Therefore, this has been extremely rewarding and on behalf of the University of Kashmir, I will say how grateful I am to the scholars who came to us from very distinguished and eminent centres, scholars like Prof. Raja Ramanna, Prof. Varadarajan and we have with us this evening Prof. Maheshwar Dayal. I am thankful to the organisers of the Seminar for asking me to preside over the Valedictory function and I am very happy that the session has concluded with the distinguished participants from all over the country. Thank you very much.

REPORT ON THE PROCEEDINGS OF THE SEMINAR AND THE RECOMMENDATIONS

G.D. WADHWA

Drafting Committee

1. Prof. M.K. Khera
2. Prof. B.D. Sharma
3. Dr. H.L. Razdan
4. Shri G.D. Wadhwa

Repportaurs

1. Prof. M.M. Bokadia
2. Prof. Y. Singh
3. Prof. Majid Husain
4. Prof. Abdul Wahid
5. Dr. H.L. Razdan
6. Shri Arun Kumar
7. Shri A.M. Mattoo
8. Dr. D.S. Parmar
9. Shri T.N. Bhat
10. Dr. M.A. Peer
11. Dr. Ashok Aima

Besides the inaugural and valedictory functions, five technical sessions were held during the course of the seminar, i.e., June 24th to 26th, 1985.

In the inaugural session Prof. Shah Manzoor Alam, the

Vice-Chancellor, while welcoming the distinguished guests gave a description of the background of how the idea of the holding the seminar was mooted in October 1984 when the Governor and Dr. Raja Ramanna were gracing the convocation of the Kashmir University and how it materialised.

Prof. Shah Manzoor Alam in his welcome address highlighed the disastrous effects on mountain ecology and agricultural productivity of the State as a result of the massive and reckless use of firewood in order to meet the energy requirements of the people. He stressed the need to cultivate an integral approach to power development and to establish an energy development corporation in the State so as to cover all sectors of energy and to coodinate and integrate their development. He also expressed the need to establish a Centre for Energy Studies in the University of Kashmir. The Governor, Shri Jagmohan, while inaugurating the seminar expressed the hope that his government will rigorously follow the recommendations of the seminar. Shri Jagmohan said that energy was the very basis of our existence and covered the entire gamut of human life. He reminded that we hardly realise the extent of damage being caused by the energy crisis. He pointed out that even a 10 per cent power cut caused a loss of Rs. 7,000 crores annually in the industrial sector alone and further stated that if we could bring down the transmission losses from the present level of 21 per cent to about 10 per cent obtainable in the developed countries there would be no energy crisis. He also showed concern about the environmental and ecological damage being caused by the consumption of huge quantities of wood.

While referring to the industrial sector, Shri Jagmohan said that although we had achieved considerable progress in the indigenous oil production, in other fields our utilization capacity (48%) was much below the prescribed norm of 58 per cent and similarly we had poor implementation capacity of plan projects.

The Governor made a couple of suggestions for the seminar to consider and ponder over in its deliberations. These related to: (i) the architectural aspect of energy, (ii) the development of techniques relating to solar energy and especially biogas which, as the Governor said, had the potential of transforming the pattern of human settlements and achieving self-contained economy at the sub-town level, (iii) conservation and management of energy, and (iv) provision of nuclear electricity to the J&K State from a central grid. Regarding the technological difficulties for the development of biogas technology for the Kashmir Valley where the temperatures are low he lamented that we as a nation have a tendency to give up quickly. He exhorted the scientists that they should instead find ways and means of overcoming the difficulties.

Shri Jagmohan said that there could be no greater resource than the inventiveness of human mind which alone could transform our liabilities into assets but the innovative spirit is undermined if social and cultural ethos of a nation generates a sense of indiscipline among its people. Therefore, what is needed is a new cultural revolution not only at the national level but also at the international level. He posed a profound question as to who would supply the necessary energy to trigger such a revolution.

At the inaugural session of the seminar Dr. Raja Ramanna, Chairman, Atomic Energy Commission, in his Keynote address spoke on the 'Problems and Prospects of Power in India' and stated that India had chosen three options for the generation of electricity, namely, (i) hydel, (ii) thermal and (iii) nuclear while the commercial feasibility of other sources of energy such as solar, biogas, etc., was yet to be established. He discussed the comparative merits and the problems associated with the three options and concluded that our national plan should strike a suitable balance of adopting an optimal mix of various options keeping in view the long-term perspective. He also emphasized the need to develop an efficient transmission and distribution system and felt the need for a grid at the national level.

With regard to the energy needs of Jammu and Kashmir State his advice was the need for the maximum exploitation of hydel power which has a potential of solving the problem of this region. With regard to new and renewable sources of energy, Dr. Ramanna said that although commercial feasibility of other sources like solar and biogas is yet to be established we should not give up merely because many a time a new scheme appears to be non-feasible. Dr. Ramanna recalled the remark of Dr. H.J. Bhabha that "no power is more expensive than no-power"

In his presidential remarks, the then Deputy Chief Minister wanted the seminar to concentrate on finding answers to the following questions:

1. Is it worthwhile to reduce the outlays for other sectors so that adequate outlays are made for power sector and whether recommendation to this effect can be made to the Planning

Commission?;

2. Whether the I.D.B.I. be allowed by the Planning Commission to invest in generation of power; and, lastly
3. Whether foreign capital should be allowed for investment in the power sector?

Shri Thakur also made a welcome offer that the State Government would be willing to finance any research project relating to power that the University may like to sponsor. He also wanted the Power Development Secretary to bring to the notice of the participants the measures taken by the State Government to bring down transmission and distribution losses and the financial constraints under which the State is working. As stated earlier, the seminar was conducted in five technical sessions. The theme of the first session was energy requirements of J&K State. The first paper on the problems and prospects of electrical energy in the J&K State by Qazi H.U. Naqash which could not be presented by him due to his absence from Srinagar gives a vivid account of the earlier history on the subject and highlights the effort the State Government is making in the development of electric power. Shri G.N. Mujoo, Dr. D.S. Parmar, Dr. Nisar Ali, Prof. Majid Husain and Dr. P. Zutshi and Shri M. Dutta presented their papers followed by a lively discussion. The following main points emerged from the session :

1. In the scheme of things the J&K Hydel Power Station will get precedence over thermal and nuclear power plants.
2. There is an urgent need to reduce the transmission and distribution losses.
3. We should try to locate and evaluate all possible projects and find resources to finance them.
4. We should try to devise ways of optimising generation from the existing power stations.
5. Ecological effects of the hydel stations should also be kept in view and a massive afforestation drive should be launched.
6. The Planning Commission of the Government of India and State Planning Department should keep in view the importance of the electric power for the economy of the State.
7. We should try to get finances from as many sources as possible.

On a point raised by Shri A.H. Mattoo, Dr. Raja Ramanna suggested that the seminar should adopt the following as resolution.

> Resolved that the Central Government be requested to allocate an equitable share to various States of the region in which the major Central Power Plant is located. In view of the policy of the Government of India that all major projects should be in the Central sector and in view of that fact that the funds which would otherwise be allocated to the States get diverted to such major power projects.

The second technical session was devoted to the conventional sources of energy. Dr. S. Varadarajan in his keynote address dwelt at length on the subject and commended the papers presented in the seminar and said that these papers represented very good collection of data, very perceptive examination of the future needs and very thorough analysis of different types of energy that could be developed in the State.

Dr. Varadarajan expressed the view that the rate of energy requirements in this State is going to increase much faster in the coming years than in the past mainly because of the population pressure and greater urbanisation, etc. Therefore, this would add to serious difficulties which were not being appreciated today. According to him there would, of course, be some compensation over the period of next five to seven years in the sense that oil prices are likely to come down considerably in the world market. But in any case oil as well as other conventional sources were not going to last for long. He further made a point that our coal production targets fixed for the turn of the century (400 million tonnes) are still very low.

Dr. Varadarajan stressed the point that enough planning and effort was not being made for the movement of coal, oil and other materials, especially from the coasts and refineries to northern India and particularly to the Jammu & Kashmir State for this would create problems of serious nature where the movement beyond the end of railway line was concerned. He suggested that to face these difficulties we must have integrated plan for the next 10 to 15 years. He advised that we should have a close look into the stocking operations of energy materials in the State keeping in view the minimum needs. Since the transportation system will go on increasingly overloaded we should therefore import only high energy coal.

Regarding hydel power Dr. Varadarajan said that there was not only a need for long-term planning but there should be an open debate on the manner of execution of hydro-projects in the State as these will be carried out in more difficult and less accessible terrains. He stressed the point that agriculture in the coming years would have to become much more intensive if it is to meet the demand of an additional 225 million people by the turn of the century and he stressed that we must produce at least 6 to 6.5 tonnes/hectare instead to the present average of only 2 tonnes/hectare.

Dr. Varadarajan also underlined the need for coordinated research in the field of energy and felt that establishment of a State Council for Science and Technology will help in this regard. He also suggested that the State Financial Corporation should be asked to fund the high risk projects dealing with research in the field of conservation of energy. The need for setting up of an Energy Advisory Board was also emphasized.

Prof. S. Manzoor Alam, the Chairman of the session, while thanking Dr. Varadarajan for delivering the Keynote address stressed the need for establishing a Centre of Energy Studies in the University of Kashmir which will also ensure greater interaction between the University and the Government. He expressed doubt whether we could use energy from the glaciers without causing tectonic problems.

Shri B.N. Kaul in his paper stated that although much work has been done on investigating various hydel projects, lack of funds is standing in the way of implementing even those which have been cleared by the Central Electricity Authority and other forums. He said that investigation of the storage hydel projects in difficult towns like Gangabal should be entrusted to specialised agencies for which accommodation needs should be met by the Government of India.

Dr. M.S. Bhat in his paper highlighted the significance of firewood to the State's economy. He also pointed out that not enough attention is being paid to fuel planning and this has contributed to denudation of forests to a dangerous extent. He suggested that effective linkage must be established in the State.

The third technical session was devoted to the nuclear power. Delivering the Keynote address Dr. P.K. Iyanger discussed the fundamentals of the nuclear power production and described various nuclear fuel cycles based on natural uranium, plutonium and thorium. Dr. Iyanger stated that the general impression that nuclear plants are dangerous was not well founded. Regarding siting nuclear power plants it was mentioned that some criteria like seismic stability, availability of minimum power load, facility for transportation of heavy and voluminous equipment have to be kept in view.

Dr. M.R. Srinivasan spoke on 'Nuclear Power' and briefly presented the scenario of nuclear electricity generation in the world. He also spoke about the Indian Nuclear Power Programme at some length and gave an account of safety standards and waste management techniques adopted in the country. As regard the energy needs in Jammu & Kashmir he did not foresee the possibility of nuclear electricity for another ten to fifteen years because of the lack of certain threshold conditions necessary for the economical functioning of a nuclear unit. Dr. Srinivasan advocated the tapping of hydel potential of the State to a maximum extent and as a short time measure suggested the possibility of using small capacity diesel generators and oil-fired thermal plants.

Dr. Krishnan submitted a paper on the 'fabrication of Nuclear Components' and could not be present to read the paper. His paper gives an account of the tremendous progress achieved in the country for the fabrication of main reactor vessel 'Calandria'. This has resulted in reducing the completion time (of the vessel) from seven years to only one year and this has led to the nuclear power becoming more competitive.

Mr. G.V. Nadkarny in his paper on 'Operational Aspects and Experience of Nuclear Power Stations' highlighted the global impact of nuclear power technology on the commercial generation of electricity. He spoke about the operational experience gained in the design, construction, commissioning and operation of various nuclear units in India which has given lot of confidence to the engineers, scientists and Indian industry to take up the challenge of nuclear tasks ahead.

During the discussion that followed it was suggested that to overcome the power crisis we could go in for a thermal power station near the railhead in Jammu. Besides, it was mentioned that if the present trend of falling petroleum prices continued it would be preferable to request the Government of India for higher allocation of kerosene and diesel to meet the requirements of energy in the State. In his remarks as the Chairman of the session Dr. M.R. Srinivasan disclosed that the cost of nuclear power in India compared favourably with the cost of the generation of thermal power.

In the fourth technical session on 'Alternate Sources of Energy in the Jammu & Kashmir State' Prof. M. Ramakrishna Rao delivered the Keynote address and spoke on the Indian scenario in solar-thermal energy systems. He mentioned that if we could trap 1/10th of the solar energy falling on India, our energy requirements could be met adequately. While underlining the paramount importance of R & D in harnessing the solar energy he stated that solar thermal energy systems using concentrating solar collectors for converting solar energy into heat energy could contribute significantly to the nation's future energy needs.

Prof. Ramakrishna Rao mentioned that during the last three years the quality of equipment for utilization of new and renewable sources of energy has been improving and the prices have been coming down drastically. During the discussions Shri M.S. Bhat stated that convenience of the consumers should also be kept in view. Prof. Rao disclosed in reply to a question that no part of the equipment is being imported except glass.

Shri Arun Kumar in his paper listed the steps which were proposed to be taken in this State for harnessing new and renewable sources of energy. Prof. M.M. Bokadia in his paper on biomass suggested that we should take advantage of China's experience of distributing biomass devices on a large scale. Dr. R.K. Pauchauri and Prof. K.L. Chopra, however, stated that in the recent past the Chinese have not been able to increase substantially the number of these devices. Dr. Pauchauri also stressed the importance of energy conservation and of the need for low cost generation of power through gasifiers, etc. It was felt that local dissemination of information is very important in this context. Prof. O.D. Thapar in his paper recommended that small hydros be designed for the lowest discharge so that unnecessary investment is avoided and these be hyberdized with other sources of energy such as biomass, solar, etc.

Prof. K.L. Chopra delivered the Keynote address in the fifth and last technical session devoted to 'Energy Management and Conservation'. He emphasized the need for reliable data on our experience in respect of the working of NRSC devices especially in Ladakh. He felt that NRSC utilization is very relevant for Jammu & Kashmir as one-third of the energy requirement is in the domestic sector. He expressed the view that after hydel power, biomass and solar energy deserve precedence in the State. He pointed out that even the richest nations like the U.S.A. and Sweden were concerned about the conservation of energy. He emphasized that there was practically no expenditure on the maintenance of photovoltaic device. He gave an idea about the economics of these devices also. He also examined the problems and limitations of NRSC devices. Prof. Chopra also underlined the importance of Green houses. Unfortunately his Keynote address could not be reproducted here.

Dr. B.N. Raina presented a paper on some possible ways to ease the situation on the energy front. He spoke on the studies in the improvement of the design of household 'chullahs'. He also spoke about the economic conversion of wood into charcoal. Dr. A. Aima presented the last paper of the seminar and he spoke on the issues in energy planning. He stressed that the resources should be distributed on a factual need basis.

On the basis of the deliberations, the main recommendations of the seminar are as follows :

1. To request the Government of India to allocate an appropriate share of power to the State from the Central Projects in operation in the Northern Region.
2. To strengthen the transmission link-up between the State and the Northern Grid as well as between Jammu Region and Valley of Kashmir.
3. The State Government should allocate higher percentage of plan resources to the energy sector than at present.
4. To establish a State Council for Science & Technology.
5. To establish an Energy Advisory Board.
6. To set up a Centre for Energy Studies in the University of Kashmir.
7. To bring down the transmission and distribution losses to the national level by taking necessary technical and administrative measures.
8. To request the Government of India to ensure construction of major hydel projects in this State on a priority basis.
9. To lay adequate emphasis on development and utilization of new and renewable sources of energy.
10. To explore the possibility of a thermal power plant near the railhead at Jammu.